# Popular Pigeon Racing

# Popular Pigeon Racing

James Martin

W. Foulsham & Co. Ltd
London . New York . Toronto . Cape Town . Sydney

W. Foulsham & Company Limited
Yeovil Road, Slough, Berkshire, SL1 4JH

ISBN 0-572-01751-0



Cover photograph from "The Racing Pigeon"



Photoset and printed in Great Britain by Rowland Phototypesetting Ltd, and St Edmundsbury Press, Bury St Edmunds, Suffolk

# CONTENTS

# A FLIGHT OF FANCY

For the past fifty years I, and thousands of others, have spent their summer afternoons gazing into the blue sky.

We are not mad, though doubtless our neighbours have often thought we were. We have simply fallen under the hypnotic spell of a sport called pigeon racing.

It may appear an unlikely pursuit with little popular appeal, but nothing could be further from the truth. In some parts of Britain the sport is so concentrated that there are almost as many pigeon lofts as houses, and it is extremely popular in other parts of Europe, America and even the Far East. Look at its attractions a little more closely and it is really not so difficult to see why it has won such wide and lasting appeal.

For instance, what other sport gives an ordinary working man the opportunity to breed his own stock, manage them as a competitive team, and experience the thrill of timing in a winner which he himself has reared and cared for from infancy?

And yet it is not the fancier who is the star of the sport. Make no mistake, in pigeon racing it is the birds who are the stars, and I, for one, hope that it always remains so.

The racing pigeon may not possess the majesty of the eagle, the speed of the hawk, or even the poetic popularity showered on its closest relative the dove, but the uncanny homing instinct which enables it to return to its owner from hundreds of miles distance has both puzzled the scientist and delighted the fancier, for it is this which makes the sport possible, and has brought enjoyment to millions over the years.

## The racing pigeon

In fact, the racing pigeon has a distinguished history, though unfortunately man has not always felt so sympathetically disposed towards it. When first domesticated it was probably thought of as a valuable food source. This was certainly the case with the Egyptians, for instance, though other cultures such as the Hindus and the classical Greeks held them in greater reverence. The Romans, though they too reared pigeons as food, also used them as an early 'telegraph' system, sending messages from city to city, and probably it was they who brought the birds to Britain.

About the same time in history, rock pigeons, which bred naturally in coastal areas all round Britain, were also being domesticated on a large scale and in some places there is evidence of man-made steps in caves, which may well have been used to allow access to birds, or at least to their eggs.

When the Normans built their castles in England it had become usual for dove and pigeon 'cotes' to be part of the plan, and as their popularity spread – again, alas, because of their culinary qualities – the richer landowners would build cotes near their grand houses. Even in these early days of domestication, the modern loft was taking shape, for soon the stone surroundings had been replaced by less expensive wooden ones, elevated above ground by the use of pillars to protect them from damp and rodents – still a necessary prerequisite for today's fancier when he plans his loft. Early cotes would be sheltered from the colder prevailing winds, and invariably faced southerly. Thousands of pigeons were raised in this manner, and even their droppings were used on farms as an excellent fertiliser.

One problem that did exist was that pigeons invariably strayed onto neighbouring farms and ate the corn. This problem still persists, though nowadays it is mainly related to the larger wild pigeons, as most domestic birds are adequately looked after by their owners. The wild pigeon is also suffering from the use of poisonous chemicals in agriculture, and these birds are becoming increasingly rare.

The pigeons seen in streets, and particularly in Trafalgar Square in London, may well make an interesting tourist attraction, but they would simply not be given house room by the serious fancier. Some may be birds lost in races, but their new diet of bread and whatever else may be fed to them, their vulnerability to disease when flocking in such numbers, and the obvious deficiencies in breeding which can never be counteracted, all combine to render their rehabilitation impossible. The racing pigeon as we know it today is in a totally different class, something which you will have already discovered if you have become at all familiar with their many qualities.

The pigeon as a messenger has been invaluable to man,

especially in wartime. During the Prussian war they carried messages from the besieged city of Paris, and more recently in the 1939–45 conflict they were extensively used by the British and Allied forces to carry vital military data and valuable information, resulting in the saving of many lives, especially among the Royal Air Force crews. Often these well-trained birds were dropped or liberated behind the enemy lines of communication, and through sheer courage and tenacity reached their home bases, accomplishing flights of 400 miles (644 kilometres) and more in atrocious conditions.

The National Pigeon Service was set up to harness these talents, with lofts at selected locations, often close to appropriate air bases. Pigeon fanciers doing their national service were entrusted to care for them and train them. Many of the birds had been donated by fanciers from all over Britain.

There are many stories told by fanciers about these winged war heroes. When a Catalena flying boat had to ditch in deplorable weather in Northern waters, a plucky little bird called White Vision struggled for nine hours in visibility of 300 yards (274 metres) to reach its loft and alert the authorities to the vessel's distress. Eleven men, who would otherwise almost certainly have died, were saved. Others who accomplished long and vital flights were Paddy, William of Orange, Ruhr Express, Flying Dutchman and Duke of Normandy.

**Pigeon racing**

If you had to pick a point when pigeon racing became a popular sport, it would coincide with the advent of the railway system which made it possible for birds to be taken in large numbers to a 'liberation point', at which they were released and left to fly back to their eager owners.

In 1897 the Royal Pigeon Racing Association, commonly referred to as the RPRA, was formed in Leeds, eventually moving to their present headquarters at The Reddings near Cheltenham in Gloucestershire. Each year they issue approximately 1,750,000 rings for young birds.

At six days old, the 'registration ring' is put on a young bird's leg, and thereafter it is easily identifiable. And such is the excellent communication between fanciers through this organisation that lost birds are often returned safe and well to their owners. The thing to do if you find a lost bird, or if one decides to have bed and breakfast at your loft, is to take a note of the ring number and report it to the RPRA headquarters, where the registration records can be checked.

Within the United Kingdom there are five racing pigeon unions; the Scottish Homing Union, Welsh Homing Union, Irish Homing Union, North of England Homing Union and the RPRA, which in itself is larger than the four others put together. For administrative purposes, the RPRA is divided into twelve regions responsible for the clubs in that area, and very strict rules under which the sport is practised are in operation. Total membership of the RPRA is a staggering 125,000 – staggering especially when one considers that each member has his own team of pigeons which may vary in strength from ten or fifteen birds, to upwards of a hundred in the larger lofts.

Racing the birds is a highly calculated business. Of course, they don't all finish at the same point though I have spoken to non-fanciers who were under the impression that they did! And because the distances they are flying are not always the same, it has been necessary to set up a system through which a winner can be fairly and irrevocably chosen. Races are decided on velocity: the pigeon flying the greatest number of yards per minute wins the race. The RPRA still calculates the distance in yards, which is then translated to metres if necessary. This is calculated by dividing the time taken to fly the distance, into the actual distance flown, with not even the smallest decimal disregarded when the mathematics are done. For instance, a bird flying 60 miles (96.5 kilometres) in exactly 60 minutes would obviously be travelling at a mile per minute (1.6 kilometres per minute), or 1,760 yards per minute (1,609 metres per minute), though that is obviously a simplified calculation.

I remember the days when a devoted club secretary, or more usually his wife, would work for hours after a race to work out the velocities of every bird, but nowadays even the smaller clubs have computers which do all the thinking for them. It goes without saying that the individual members of the clubs have had the precise locations of their lofts determined, and this too is done in detail using the exact latitude and longitude.

When you send a pigeon to race, it will be basketed, and a rubber ring placed on its leg. When it returns to your loft, the ring is removed and placed in a highly accurate timing clock, which then registers the day, hour, minute and exact second of timing in. Then the velocity is worked out and hopefully your bird's performance will be better than anyone else's in the club. If the bird has done particularly well, it may win a regional prize, or even a national one, in which case your achievement, and your cash winnings, will be substantial.

## Pigeon fanciers

More of actual racing later, but what of the people who have these birds in their care? I think that almost without exception, the owners of pigeons are extremely responsible people with great feelings of affection for their birds. They will not usually allow them to race unless they are physically well prepared, for why would they allow a bird that they have reared, fed, watered and cared for, to set off on a journey which it has little chance of completing? Unfortunately there are hazards, ranging from ordinary telephone wires which tragically cost the lives of many pigeons, to the unpredictability of the weather, though every effort is made to ensure that conditions will be at their best when birds are liberated.

Silhouetted against the evening sky, these birds have been frozen as they take their nightly exercise.

It is not the cheapest of hobbies, what with feeding, providing accommodation and equipment and paying club and race fees, but it is a rewarding one and I think the special relationship between bird and owner is what makes it so. There are many more 'pigeon racing widows' around the country than many people would expect – this is undoubtedly why so many wives have taken an interest in the sport, and the race results now refer to a 'Mr and Mrs' partnership just as often as they do to a father and son team.

Interestingly too, more and more professional people, like doctors and lawyers, are taking up the sport, and I like to think of it as a sport for people in every walk of life, for each can devote the time, energy and money appropriate to their own resources. Not only is Her Majesty Queen Elizabeth the patron of the Royal Pigeon Racing Association, she also has a racing loft and birds at her Sandringham estate where they are trained by loft manager Len Rush. Other well-known personalities also enjoy the sport – among them racehorse trainer Sir Gordon Richards, and international footballer Gerry Francis, who races a strong team in the London area.

Pigeon fancying has many famous devotees, among them former England football captain Gerry Francis, seen here admiring one of his birds.

I would never advise a potential fancier to take up the sport for commercial purposes, but considerable sums of money can be won in the Derby and National races. Some of these have added sponsorship, and a well-pooled pigeon can win thousands of pounds for its owner in a single race. The provision of feedstuffs, loft accessories and health aids for birds, as well as the breeding of pigeons, has seen the sport develop into a multi-million pound industry far removed from its humble beginnings.

But win or lose, I am sure that if you decide to take up the sport, countless hours of enjoyment are in store. Many problems lie ahead too, but hopefully I can help you solve some of them with the tips and information I shall endeavour to pass on in the following pages.

Pigeon racing is a fascinating and rewarding hobby, not least because of the special relationship between a fancier and his birds.

# THE LOFT

If you have decided to enter this absorbing and intriguing sport seriously, an absolute necessity is adequate accommodation for your future stock.

One of the most common dangers to be avoided by new fanciers is the natural tendency to imitate the elaborate loft designs of leading fanciers. Even these keen fanciers would admit that this adds little, if anything, to the success or racing performances of individual birds. Under no circumstances should this be taken as an invitation to keep birds in delapidated conditions. I am only emphasising that a small, well-maintained loft, regularly cleaned and painted, is just as likely to produce the future champion, as is a palatial building with all the modern conveniences.

Experience has shown that the small back yard or garden loft, well managed by an enthusiastic fancier, has, largely because so few birds are kept, allowed the fancier to treat each inmate as an individual, rather than looking on his birds as a collective team, where the owners can be tempted to concentrate on quantity rather than quality. Birds which receive individual attention are much more likely to remain content and perform to the best of their ability, both as racers and breeding stock.

**Climate and weather**

Having visited fanciers and lofts in many parts of the British Isles and Canada, I have discovered that no specific design is common to all fanciers. With the variable and seasonal weather changes which are experienced in the British Isles, parts of Canada and many European countries, a loft with a partially open or dowelled front suitable for warmer and drier areas would not suit the colder and wetter weather in more northerly regions. You must therefore consider the design of your loft carefully in relation to your location, as the comfort of your birds is vitally important.

## Costs

Pigeons that inherit a love for their home will race with success either to a very modest structure, simply constructed by a novice handyman, or to a more elaborate building, perhaps designed by an architect and complete with running water on tap. In my opinion both lofts will have their successes and failures for I am fully convinced that cared-for pigeons, well-trained and possessing the will to win, will not think of grandeur, nor will they care if the loft is painted in their favourite colours! Their greatest desire after a long flight will be to reach food and water and return to their daily chores. This is especially true if the bird is sent off to race while 'clocking' eggs, the term used when birds are looking after eggs, or if feeding youngsters. It wants to get back to its eggs, or to its offspring or its mate, and therefore if it is happy in its loft, this responsible bird will be in no mood to give them up lightly.

Perhaps your first essential consideration would therefore be the financial aspect of housing birds. Decide how much cash you can afford, remembering that the provision of the loft is only an initial outlay and that to follow you will be faced with increasing feeding costs, club subscriptions, the purchase of a timing clock, hampers, nest bowls and numerous other essentials. Don't be alarmed, though, if all this adds up to a considerable outlay. You may well be able to pick up a few bargains from helpful fanciers, not least in the area of the loft, where fanciers often chop and change and therefore want to sell their old loft. The thing to remember is to be realistic about what you can afford.

## Permission

Another point to remember is that you should check with your local planning or housing authority. Find out if it is possible to obtain permission for a loft if you live in council accommodation. If you rent a house from another source the same applies. Other fanciers living in the same area may well be able to guide you, and approval should be less difficult to obtain if there are other lofts nearby. But the last thing you want to do is build up your own hopes and invest in a loft only to find that it poses problems with the authorities.

I think it is also important to have a chat with your neighbours as they may well be alarmed at suddenly finding a sizeable, permanent structure next door. It is only common courtesy to do so, but remember to tell them that pigeons are intelligent and clean birds, unlikely to trouble them to any unreasonable extent.

## Buying a loft

If you want to buy a second-hand loft, and as I said these are often available and can be excellent value for money, you are most likely to find someone in your own club who has a loft to sell. Otherwise you may have to consider buying a new loft, in which case the first place to look is the fancy press.

Pigeon fanciers in Britain are fortunate in having at their disposal two excellent fancy press weekly publications, *The Racing Pigeon* and *British Homing World*, to which I have been pleased to contribute items for many years, and they adequately

If you want to buy or make a pigeon loft, be careful to plan the design and site very carefully as you will not be able to make subsequent changes very easily. This loft uses a dowelled front, which is both practical and attractive.

cover the new developments in the sport. Equally popular with the readers, though, are the classified columns, for there the fancier will find advertisements for many reputable firms who specialise in the manufacture of lofts and equipment. On request, most of these companies will send a brochure or catalogue giving detailed information on many different loft designs. Many firms will also be pleased to give an estimate for a loft built to your own design and specification. If you want to get into the sport quickly, and the money is available, these lofts undoubtedly offer good value and in my experience the workmanship is usually first class. If you are new to the sport, however, do take expert advice on the design of your loft, as it will be difficult to change if experience proves that the design was not of the best.

**Building a loft**

If you are determined to 'go it alone' and build your own loft, let me offer some advice. It is vital that the building is dry and well ventilated, but free from draughts, with perches and nest boxes situated well away from any possible source of damp, as nothing else upsets the constitution and health of racing pigeons as much as continued dampness. In this respect loft design and position should be studied in detail. The best position for the loft will often be determined by the immediate surroundings, but on no account should the loft face to the east because of the cold winds blowing from that direction. Second only to dampness, easterly winds are greatly disliked by pigeons and the proof of this can be seen in their below-par performances on race days when the line of flight faces easterly and the wind is blowing.

Also remember that it must be easy to clean – hygiene is essential – and that you will be spending a considerable amount of time there, so it must be comfortable for you as well as for your birds. Make sure the roof is high enough for you to stand upright, for example, and that the nest boxes are easily accessible.

Assuming you are confident and competent enough to build your own loft, make sure that you take the utmost care in the workmanship. It will be worth it in the long run, for a carefully made and well-maintained loft can last a lifetime. For convenience the loft can easily be made in sections, not too large or awkward to handle, using nuts, bolts and washers in the process.

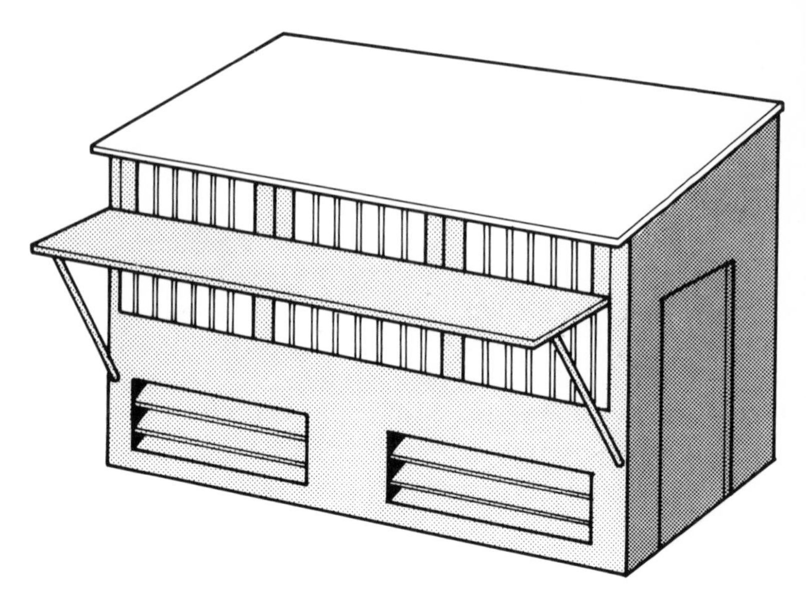

The front of the loft should allow for good ventilation so incorporate home-made or bought ventilators to ensure a flow of clean air into the loft. Unhealthy air will escape through the roof holes along the rear of the loft.

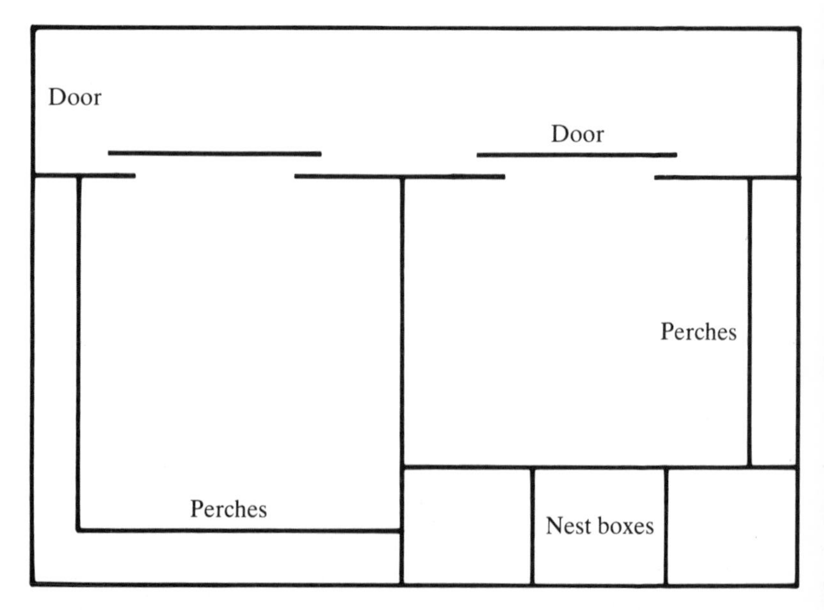


This loft plan shows a typical loft using the open door trapping method. When you plan the loft, remember that you must be able to move around it comfortably and be able to clean it regularly.

My earliest recollections of pigeon lofts were of those of entirely wooden structure, and even today there is no better material for the construction of a warm and comfortable loft, properly designed to meet the individual fancier's particular requirements. A sensible and adequate loft for the fancier who intends racing birds to his garden or back yard would measure roughtly 4 yards (3.6 metres) long by 2 yards (1.8 metres) deep, sufficient to accommodate ten to twelve pairs of old birds and fifteen to twenty youngsters.

Ideally, the floor should be raised at least 12 inches (30 cm) above ground level to ensure that fresh air circulates underneath. This will help to keep the floorboards dry and free from rot. Ideal for this are block or brick piers which have the added advantage of making a strong base and foundation for the entire structure.

Start by making the floor which should have a smooth and even surface for cleaning purposes. Choose a strong timber – I would advise 3 x 1½ inch (76 x 38 mm) sections, or even 3 x 2 inch (76 x 50 mm) for the floor joists or battens, spaced at 15 or 16 inch ( 38 or 40 cm) centres, sheeted or covered by ¾ inch (19 mm) or 1 inch (25 mm) tongued and grooved flooring. The entire flooring should be treated with creosote or a similar wood preservative. If funds are available, the floor can be covered with ¼ inch (6 mm) flat asbestos sheeting, but if you decide to do this make sure that you use proper nails or screws to ensure that the heads are not protruding. If screws are your choice, these should be countersunk. Any good hardware shop will have a supply well suited to this purpose.

It is better to tackle the back section in two 2-yard (1.8-metre) halves. These, together with the two end sections – one of which must of course contain the doorway – can be constructed from ½ inch (12 mm) or ¾ inch (19 mm) exterior plywood sheets which cut down both on the amount of work needed and the time involved. These do an adequate job if you make and measure the framework to suit.

The front I favour is one that allows a free and plentiful flow of air, and therefore I would recommend the provision of louvres or ventilators. As your loft is, for construction purposes, divided into two equal sections, place the ventilators in either half as near as possible to floor level. If you are not particularly skilled, these can be purchased independently, or perhaps you may know a skilled tradesman who would be willing to make them for you as there is a fair amount of detail involved. The inside of the ventilators should be covered by fine wire mesh to keep out possible intruders, including the disease-carrying house sparrow and small rodents.

As it is also vital to allow foul, unhealthy air to escape, a ventilation system must be incorporated. It is not advisable to interfere with the nest boxes and perches, which will be situated at the rear, so the best remedy is to allow a space of at least 3 inches (76 mm) above the top perch, with drilled roof holes of about 1 inch (25 mm) in diameter, spaced every 2 or 3 inches (50 or 76 mm) along the length of the rear section. If, however, you prefer corrugated asbestos sheeting for the roof instead of a wooden, felted roof, this will be unnecessary because of the ridging on the asbestos sheets.

The roof should have at least a 12 inch (30 cm) overlap both front and back, and not less than 6 inches (15 cm) at the sides, which will help prevent blowing rain from getting inside the loft. Again what you want to avoid is that old enemy, damp.

Although asbestos sheets are quite easy and simple to put on with approved galvanised nails and washers, and are preferable to the wooden, felted roof, they have a tendency, like corrugated zinc, to 'sweat' and condense in frosty conditions and in extremely wet weather. I have seen lofts where this has been successfully overcome by using reinforced roofing felt or thick plastic sheeting nailed down tightly before the asbestos is fitted. This should be overlapped in the opposite direction to the fall of the roof.

### Trapping

There are various ways in which pigeons can be 'trapped', the name commonly used for admitting birds to the loft after a flight. This is vitally important, of course, on race days, and so it is well worth considering carefully.

The most popular method is the verandah or outshot, which can be made in many different designs to suit the individual fancier's requirements. Have this soundly made, with the framework morticed and tenoned, rather than just roughly nailed together.

Dowles, I think, look very attractive. These should be used

in the lower section, with glass or perspex in the top, except where the birds enter the outshot, where a small door hinged to the top should open inwards. The boards on which the birds alight should have a good overhang to shelter the verandah floor. It will also be necessary to fit a small door in the bottom half to let the birds out.

The benefit of having a verandah is that when the young birds are ready for the first view of the outside world, they can see the surroundings before venturing out onto the roof. This should help diminish the danger of the much dreaded fly-away, when birds leave the loft for the first time, never to return.

Another common and quite satisfactory method used for trapping, and one that has the added advantage of being easily constructed, is the landing-board, which should be made strong and firm. The platform should not be less than 18 inches (46 cm) wide, and it should be supported by brackets made of 3 x 1½ inch (76 x 38 mm) cleaned and planed timber. The flat asbestos is ideal for putting on top of the boards and again, if properly fitted, will ensure that the landing-board is always dry.

A slight fall in the outward edge helps rain escape. A space of 8 to 12 inches (20 to 30 cm) should be allowed between the roof and the landing-board. This will entail fitting a door hinged to the roof framing, which will have to be opened to allow the birds to enter and leave when they are exercising and training. The only drawback I find with this method is that pigeons, especially youngsters, are hesitant about the drop from the landing-board to the loft floor.

### Open door trapping

This method of trapping seems to be a favourite with many fanciers especially in the Midlands of England, and is certainly an easy one to operate. If you keep your birds near your home, and they are 'under your eye' throughout the day, then the open door comes into its own, and if your loft position is suitable it would be hard to improve upon this most basic of methods. As well as having a large entrance area, the open door also provides a large exit which can be very important, because if the exit is too small – and this is a point to bear in mind regardless of the trapping method used – there is always the danger of injuries to birds in the mad scramble to take their exercise.

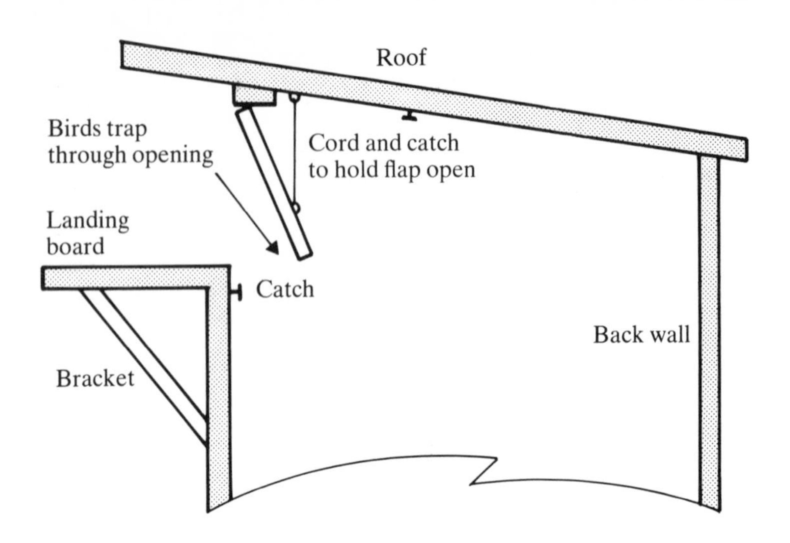


The landing board and trap is a common method of trapping pigeons.

To avoid loose birds getting out when the racers are expected, the compartment to which the birds are trapped should be vacant. Fresh birds leaving the loft could well keep the race bird circling, and valuable time can be lost.

**Dividing the loft**

The beginner should divide his loft into two separate compartments, one for the adult birds, the other for the youngsters. This also allows the sexes to be kept apart before mating. When the breeding season arrives, the young birds can then be transferred to their own section once they are ready to be weaned from their parents. The division should be as near the centre of the loft as is possible, with a door for access between the compartments.

**Nest boxes**

One section needs to be fitted out with nest boxes, and of these there are many variations and sizes to choose from. Although many lofts use the back of the loft to form part of the box, thereby cutting down on the work by only needing to provide the sides and floor, I prefer to see them in separate units of three or six so that at the end of breeding, or when maintaining or painting the loft, they can be removed easily. Irrespective of

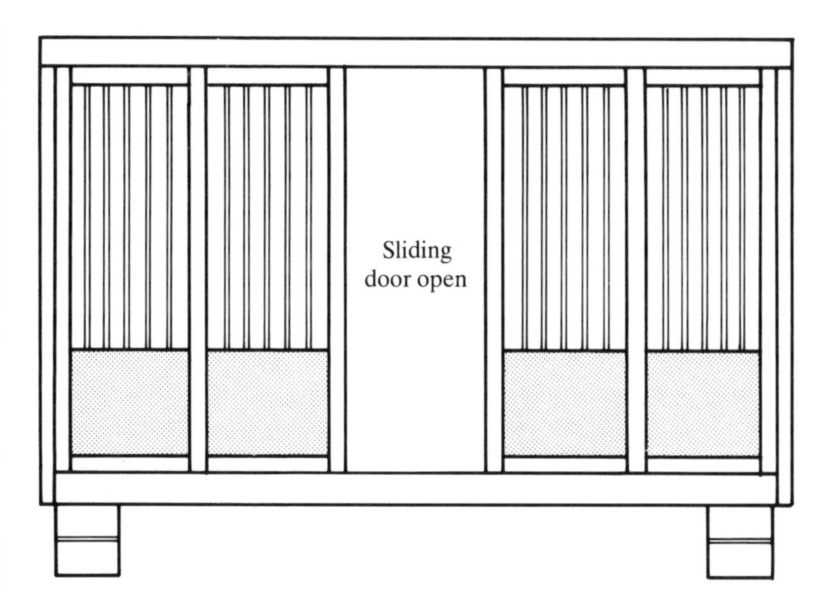


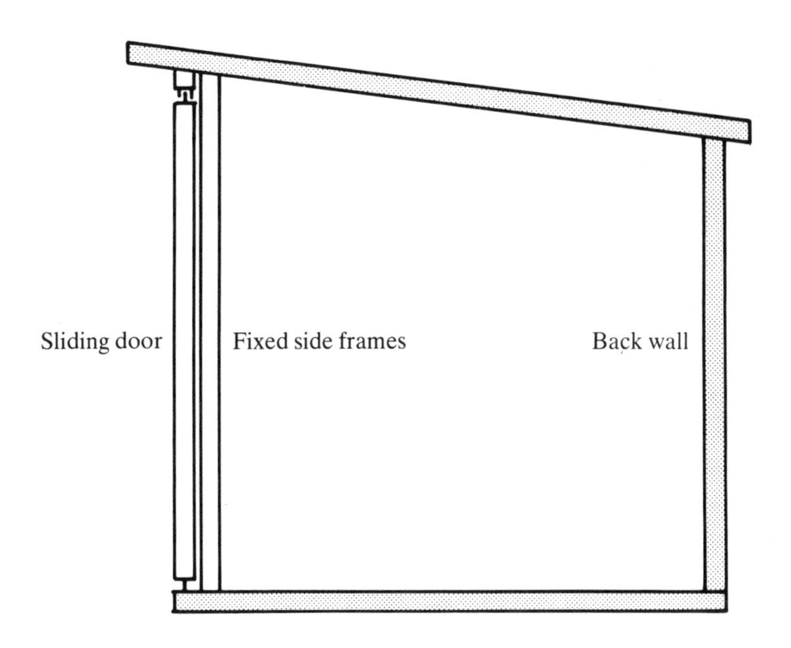


The open door trapping method is simple and effective, and particularly popular with fanciers in the Midlands of England.

what type or size you decide upon, ensure that the bottom of the nest boxes is raised 9 to 12 inches (22 to 30 cm) above the floor. This has many advantages, especially when it comes to cleaning the loft, and the space can sometimes come to the rescue of a youngster that has been unfortunately 'scalped' by some of the older pigeons.

The measurements of nest boxes have no set standard but they should of course be large enough to provide comfortable surroundings for the birds, as they will be occupying their own little matrimonial homes for more than six months of the year. A good, roomy nest box should be about 24 inches (60 cm) long, 18 inches (46 cm) from back to front and 15 inches (38 cm) high, with nest front fitted. The nest box floor should extend another 5 inches (12 cm) to allow the birds to perch if they wish, and the entire thing can be made with a removable bottom, which I find useful when wanting to apply a new coat of emulsion.

There are many common designs for the fronts of nest boxes and, like the nesting boxes themselves, these can be purchased ready-made, but as the standard size of manufactured nest boxes and fronts is often smaller than I like, I prefer to rig up my own. The ready-made box on the market is usually made of either ½ inch (12 mm) chipboard or exterior plywood, with the fronts dowelled, and a single perch in the middle allowing the birds to pass in and out. The perch is also useful if you wish to fasten the pairs up. Some firms are now making fronts with plastic-coated wire instead of dowels. Nest fronts can be fitted to the boxes by using turn-buttons which allow the entire front to be removed for cleaning.

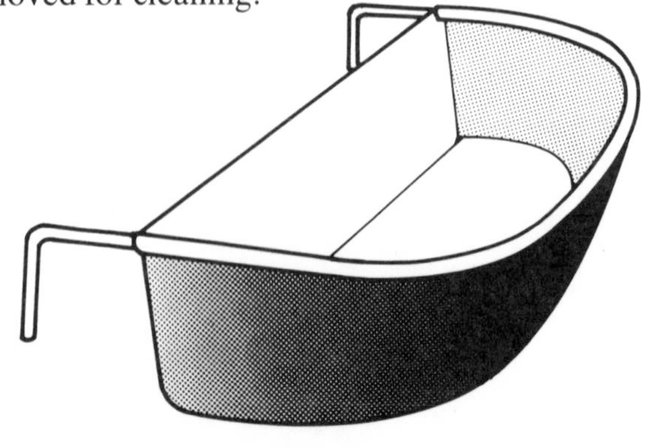

A nest box cup fits into the box so that you can supply fresh food and water.

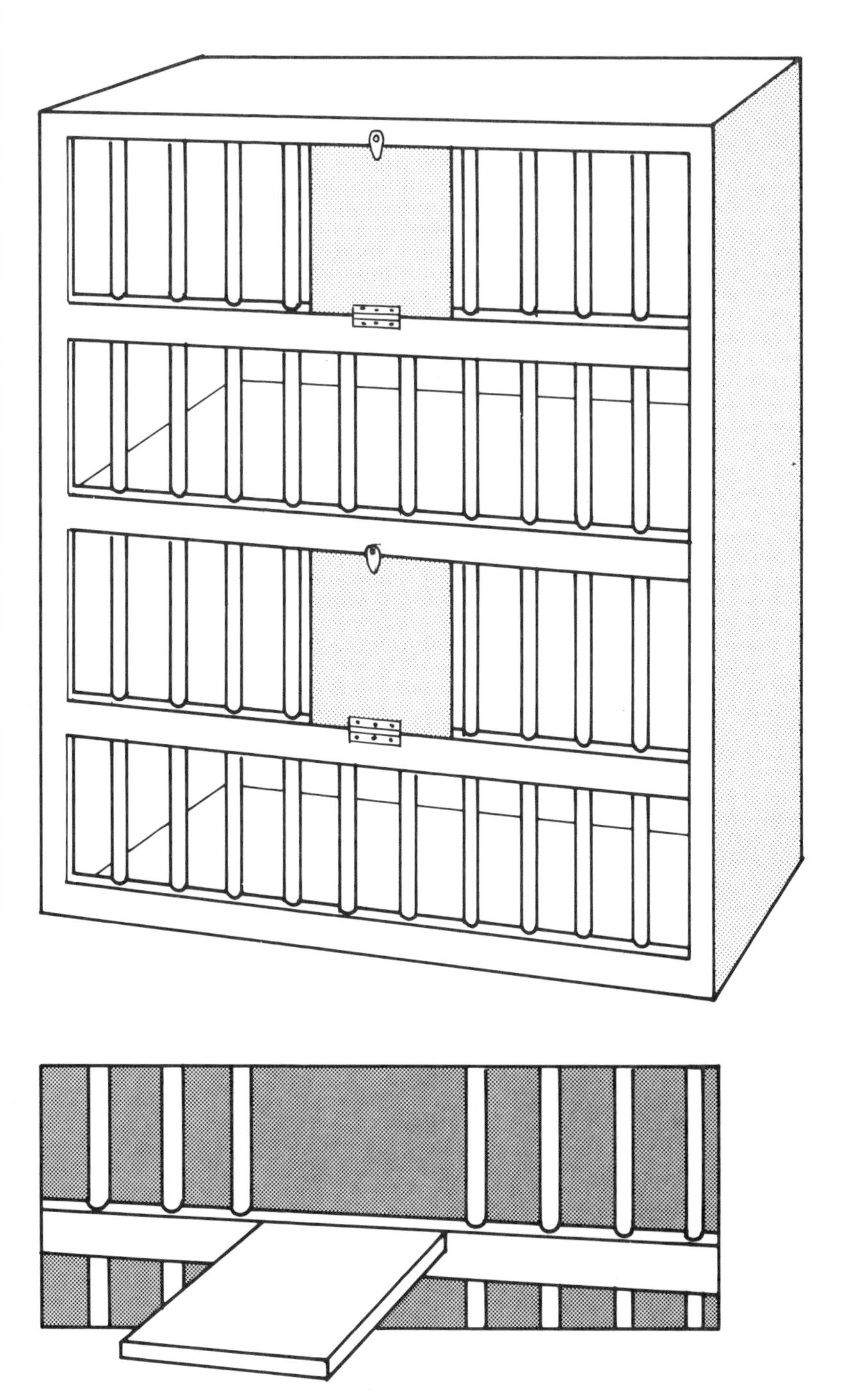

Nest boxes should be roomy, and should include a perch when the door is open.

These sizes, dimensions and types of front are for birds on the natural system, but with the ever-increasing number of fanciers now flying birds on the 'widower' system, different designs and sizes will be required. This is explained more fully on page 72.

## Perches

When the breeding section is fitted out with nest boxes, only the young bird compartment will have perches, most of which fit into two common categories. Most fanciers favour the box perch in preference to the 'inverted v', or saddle perch. The main complaint I have about nearly all box perches is that they are not deep enough – usually only 4 to 5 inches (10 to 12 cm) – and pigeons occupying the lower perches are likely to have their feathers, and particularly their tails, soiled with droppings from birds on the upper rows. The last set I made were a full 7 inches (18 cm) from back to front, and anything up to 12 inches (30 cm) would be even better. Some fanciers may argue that the wider perch will collect droppings from the bird installed there, but I would much rather clean the perch, than a soiled or dirty-feathered pigeon.

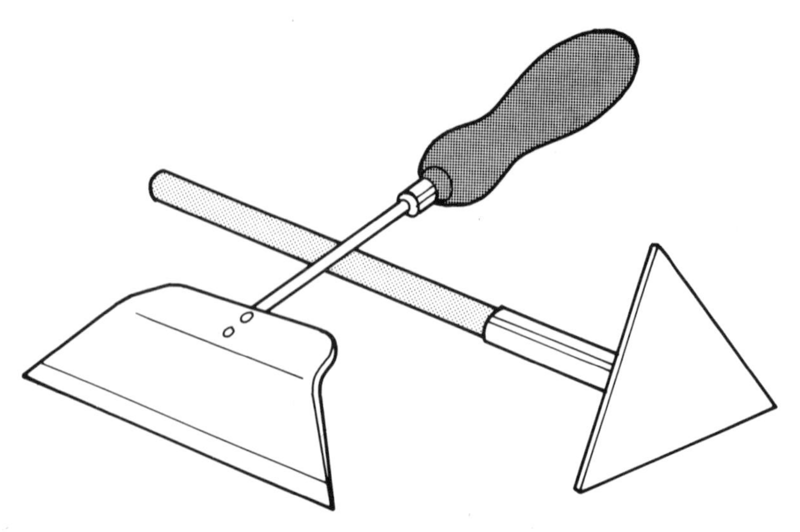

Remember that you must be able to keep your loft clean, so bear this in mind when constructing it. There are a variety of scrapers available for cleaning the loft.

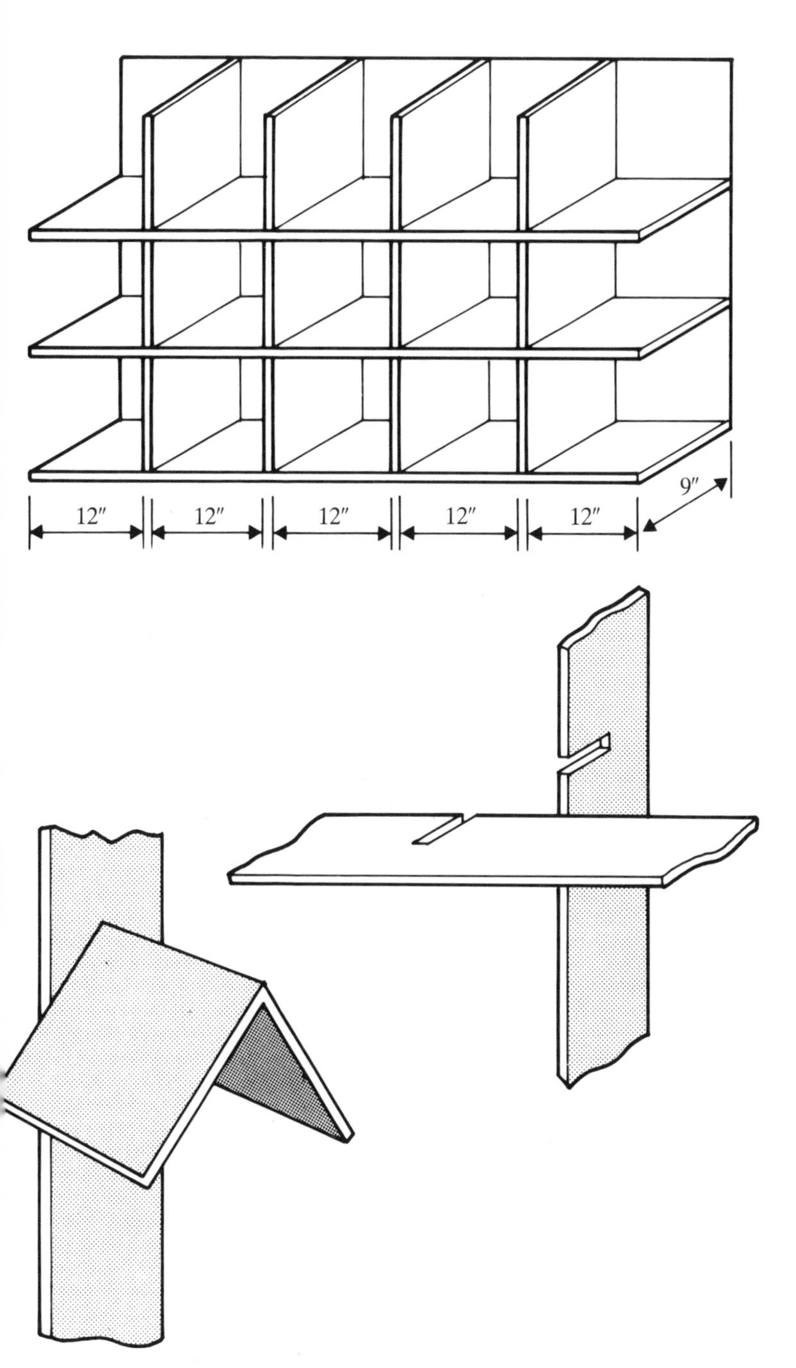


The two main types of perch are the box and the saddle or 'inverted v' perch. Whichever you choose, make sure they are deep enough to prevent the birds soiling the bird below.

### Feeding and drinking bowls

When the loft has been erected, complete with division, nest boxes and perches, any birds you require will need to be fed and watered, so some kind of receptacle will be needed for this purpose. It is essential that both are always kept clean and hygienic. When I first became a fancier and money was in short supply, and indeed when water on tap was something of a luxury in our homes, my drinking system was nothing more than a jampot with bricks set either side of it for stability. The problem with open vessels is that the water very quickly becomes contaminated. Nowadays ideal covered water fountains can be easily and relatively cheaply purchased through the many fancy suppliers, and most good pet shops also stock them. They are certainly much more practical than my old pre-war jampots!

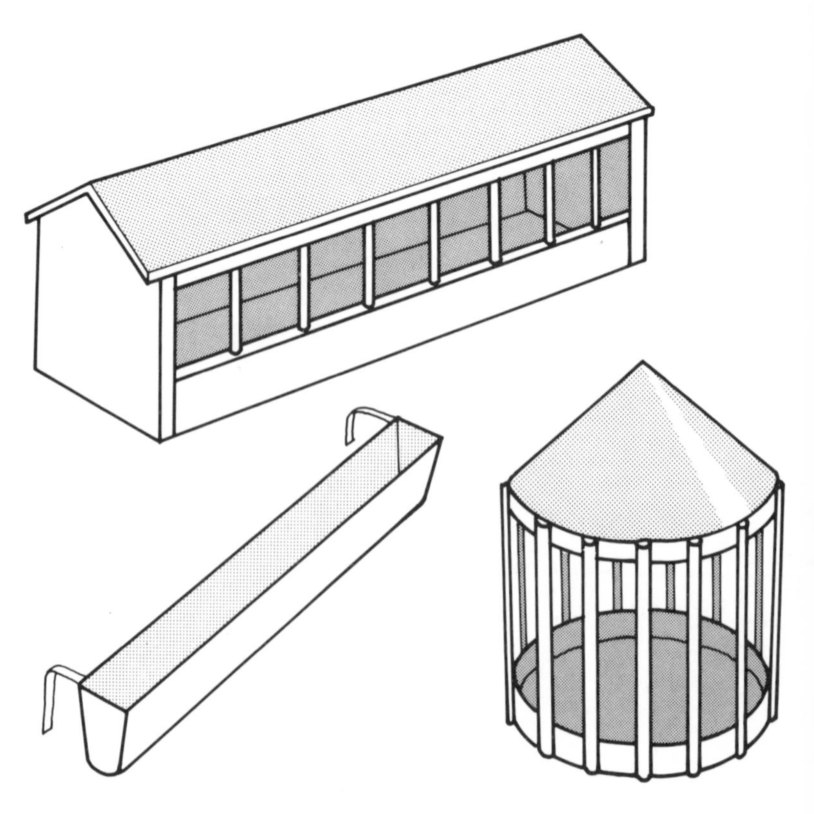

Feeding and drinking containers now come in many shapes and sizes,so you can choose whatever suits your loft. It is best to use a covered water fountain, however, to avoid contamination.

# BIRDS AND STRAINS

The novice fancier will appreciate early in his career that, regardless of the source from which he acquires his foundation stock, only a small percentage will in turn breed top class progeny. Nor has any fancier yet been gifted with an abundance of top class birds with outstanding breeding potential to such an extent that he is in a position to sell them either to a novice or anyone else, irrespective of how much cash is offered. Many fanciers have realised, sometimes to their considerable cost, that money does not buy success, and that expensive stock is not always the best.

A common mistake not only committed by the beginner but also by the experienced fancier is to be influenced by the exorbitantly high prices that are paid for so-called pure strains, backed up by pages of pedigrees bred from unraced ancestors that in many cases never left their breeding pens.

It is a good idea to lift and handle your birds, especially new stock, and accustom them to your presence around the loft.

## Buying foundation stock

Once in a position to accommodate your birds, it is necessary to decide if you are going to buy a few pairs of old pigeons for stock and breeding purposes, or acquire a number of young birds, which in due course can either be raced or retained for stock. These decisions will ultimately depend on the available facilities, the loft accommodation, and your financial limits, which should not be exceeded if you are to enjoy the sport to the full.

It is only to be expected that few fanciers will want to let you have a proven stock pair, but if you are lucky, and if you are prepared to shop around, a suitable pair could be obtained at a clearance sale. This would be an ideal start, but do avoid buying old and worn-out stock pigeons, for even if these birds are breeding at their present loft, where possibly they themselves have been reared, a change to new surroundings will curb their tendency to continue breeding strong and healthy young stock.

The best method of starting your stock, and one that I would strongly recommend, is to approach a reputable fancier in your locality, one who has a distinct family of racing pigeons. State your needs, and be prepared to place your trust in his hands. Remember that his experience will enable him to select birds with potential. It will possibly mean paying a little more than average but I believe the extra expense in this case is well worthwhile.

If possible obtain birds that come from the best long-distance winners, and if you choose a genuine fancier from whom to purchase, you can be sure that all the details with regard to performances and pedigrees will be authentic.

## Settling into the new loft

The first thing to check when you have bought stock is that they are settling in to their new surroundings. Make sure that all the birds are eating and drinking, as some young pigeons when taken away or weaned from their parents will refuse food for a day or two.

If the young were well reared, they should be eating solid food on their own account, but it is advisable and well worth the trouble to lift and handle them for the first few days to feel for food in their crop. If you notice one that appears a little backward compared to the rest in feeding, dip its beak in the water fountain to encourage drinking. When handling, take the opportunity to give them a thorough dusting of insect powder. Any of the propriety brands on the market will do, and at the same time examine their beaks and throats for possible traces of infectious diseases such as canker. Ideally this should be done before purchasing but it is unlikely that any reliable fancier would not have done so already on your behalf.

### Eating and drinking

A little bit of biological information on eating and drinking. Birds digest their food in two stages once it has reached the stomach. Firstly the food is attacked by enzymes, and then the gizzard, which has very strong muscles and sharp little ridges, acts as a grinding system to help break up the hardest of grains and seeds. Don't be alarmed to see your bird pecking at small stones, for it does this deliberately to make the pigeon's interior 'mill' even more effective. I have seen some fanciers alarmed when their birds are eating small stones, believing they are doing so in error and that the stones will be harmful, but this is how the pigeon can best digest the hard grains on which it thrives.

Pigeons have an extremely unusual drinking system, most unlike the majority of other birds. They actually suck up water without needing to raise their heads from the source, taking long draughts which will visibly lessen the supply at hand if kept in a small receptacle.

### Training

As soon as the birds are obviously settled in to their new loft, they should be allowed to see their outside surroundings as soon as possible. This is where the trap or verandah described on page 20 becomes a real asset, as the new birds will spend a good deal of time sitting around, summing up the outside world, in complete safety from any outside dangers. At this stage they should be accustomed to regular feeding, preferably receiving a light feed in the early part of the day and a full crop at night, and at this stage I would familiarise them with the rattle and noise of the corn tin, for once they start to bunch and fly, they must be educated to trap quickly and not sit around on the roof tops once fully on the wing. All too many races have been lost by birds failing to trap quickly enough, losing valuable seconds and even minutes after having done the hard work of regaining their lofts.

At all times avoid rough handling of birds. To give them their liberty outside the loft for the first time, open the verandah doors and leave them alone, preferably in the late evening. Eventually they will walk out, probably making their way to the loft roof, and perhaps even your house if situated nearby. Don't take the risk of catching them and putting them on the roof for their initial outing, as this can naturally startle and frighten them, and if their wings are advanced, they may fly off and be lost for ever.

This is a stage when you need to be patient, like a father waiting for the baby's first step, but rest assured that step will come. Don't force the birds to fly or exercise before they become accustomed to their outside surroundings. After tentative attempts to find their wings, they may even have a short fly around sooner than you would anticipate – but please don't force them, even if they do appear reluctant.

It is a memorable day for the fancier when his young pigeons eventually head off to scour the countryside for the first time. You may have some anxious moments waiting for them to return, then they will suddenly appear out of the blue to return to their loft.

The observant fancier will notice a leader among the pack. Invariably it will be first out, first on the roof and, in most cases, first to take off. When they have been out for an hour or so, and irrespective of whether or not they have flown, they should be whistled back into the loft, another task which calls for patience in the early days. Educate them to enter by the same means each time.

You will notice that when the youngsters start flying, it will be in short, sharp and playful sprints, not lasting for more than a few minutes at a time. As time goes by they should group into tight little batches, similar to the way you will have seen old birds training or exercising around a loft. They should by now be eagerly looking forward to leaving the loft for their twice-daily exercise, and for the first few flights the youngsters will remain in close proximity to the loft.

Eventually on some memorable occasion they will 'head off', and start what is commonly known as raking or scouring the countryside, and may be away from the loft for an hour or more. Anxious moments these, but just when you are becoming alarmed and wondering if you have seen the last of them, out of the blue they will suddenly appear, to circle round and drop on the loft. This can be an exhilarating development, and one which even the older fanciers enjoy year after year.

When safely back in their perches, count the birds and make sure that the entire kit has returned. As this was their first flight of any duration, it is possible that one or two may look jaded and tired, but after food and drink, and a good night's rest, they should be fit enough for the following morning's exercise.

**Birds in flight**

By now you will be able to identify easily the unmistakable pattern of a bird in flight, something which is essential for race days. The racing pigeon is, in fact, a tremendously effective flyer from a technical point of view. They are real masters of the airways, with wings that flap quickly and give them their special ability to fly at high speeds over long distances, something that few other birds share.

Pigeons have a very direct means of beginning flight and they are also expert at landing, extending their forewings as they descend to slow them down, using their tail as an air brake and at

The unmistakable patterns of birds in flight.

the same time to keep themselves steady. Indeed there is no mistaking the racing pigeon in flight, for it really does bolt through the air as if sprinting for a short distance, even though it may be on a long flight. It will of course look more leisurely and relaxed when circling the loft, but when it comes to getting from A to B, and regaining its home loft, the pigeon flies both purposefully and powerfully. Its power and speed come from its large flight muscles, long flight feathers on the wings and light air-filled bones.

The question of racing these first youngsters is a difficult one for new fanciers, eager to experience the thrill of competitive flying, but one thing that should not be overlooked in any case is proper training. In this chapter I have outlined the elementary procedures for the management of young birds, but detailed training will be covered in 'Training and racing young birds' on pages 42–56.

## Strains

Fifty years in our fascinating sport have for me brought an almost unbelievable number of changes in the strains of pigeons, especially in the past few years. Only on rare and infrequent occasions do we hear nowadays of the old English and Continental families such as the Osmans, Tofts, Logans, Gurnays, Hansennes, Barkers, and many more too numerous to mention, with which we were once so familiar.

The founders of these nostalgic strains have all passed on, and more recently we had the Kilpatricks, Warringtons, Marriotts and Lulhams, but during the last decade even these have been replaced by the more fashionable and expensive families, mainly of Belgian origin. Many of these are described as pure this and pure that, but to me this has always seemed a rather optimistic if not fanciful notion, for the question I pose is this: where have all the previously-mentioned families disappeared to, for they seem to have vanished off the face of the earth overnight? I am fully convinced, and relieved in the knowledge that many older fanciers are thinking in the same terms as myself, that many of these long-established strains, and others, were donated to fanciers in the occupied countries during the last war and have been used in the creation of these so-called pure families, irrespective of the strain or the name of the breeder.

In conversation with a celebrated fancier on the various aspects of pigeon racing, I was interested to hear him relate that in his opinion successful lofts did not own all the good pigeons, nor the less successful own all the poor ones. I would take this a logical step forward and apply it also to strains and families of pigeons.

There are good and bad in all strains, and in any case the good ones are in the minority, which makes the choice of strains that much more difficult for all fanciers, experienced and inexperienced alike. Some years ago I purchased pigeons from a long-distance loft, from an area miles from my own home, and in his words of advice the fancier told me to be patient with the new introductions for at least three years, claiming that the change of environment and climate could retard the immediate success of these pigeons. This is a theory I fully endorse. Some will accomplish outstanding performances flying in a particular

A Red Chequer.

A Blue Chequer.

direction and to a particular area, and will fail miserably when tested to a different locality. Both the route negotiated and the management of the birds will be of vital importance, and so strains that have raced consistently and successfully in your own area are the best buy. If they are of an inbred family, the choice of strain is irrelevant. Don't worry about pedigree, as good pigeons produce their own pedigree.

**Improving your stock**

As many fanciers have learned from experience, the business of mixing and matching birds for mating in order to produce good performers is something of a lottery, but if you are dedicated enough, and careful in your observation, results will come sooner or later. Good birds, and potential winners, can be ruined by bad management, just as good management may bring you better results than might be expected from the pedigree of your stock.

This Blue Chequer has a distinctive white flight.

No strain can be properly maintained without the owner practising in-breeding, which means mating up birds of the same bloodline, as close as sire to daughter, mother to son, or brother to sister. Line breeding is on the same principle, with birds again paired to close relatives.

The novice will, after a time, and by personal observation of his stud, eventually get to know what he requires, and should then avoid any change in management or system until he has had a reasonable period to either prove or disprove his stock. There should be a 'natural selection' process in your stock management through which pigeons not up to scratch, or not suited by your system, are lost. Yet in this complicated and sometimes frustrating search for success, it is always necessary to apply logic. If you have a system which loses more birds than you can afford, then change the system, or change the strain, but at least do something positive rather than sitting back and awaiting success that is obviously not going to come in those circumstances.

You may wish to concentrate on winning short races with young birds, or long events with more mature ones, and so obviously the fancier who wants the former will have a vastly different stock to the one who wants to win old bird events, but many mistakes will be made in both cases before your knowledge and understanding grow to such an extent that you can make difficult decisions wisely and profitably. This may often mean giving up after a long and unhappy trial with a particular strain or system, but it may be necessary nonetheless. There is a marked difference between the person who simply keeps and feeds pigeons, sending them off for the occasional race, and the racing manager who is constantly logging the performances, successful or otherwise, of his racing team in his bid to hit on a winning formula. Once your birds have proved that they are of top class quality, be very careful regarding any new introduction of fresh blood. Be sure of the newcomer's origins and, above all, keep a very strict eye on the new imports.

**The wing theory**

If you have visited many lofts, or have accompanied experienced fanciers when they were doing so, you will be familiar with the pigeon fancying custom of vetting and handling birds as

A Blue Barred bird with white flight.

There is no mistaking the markings of the 'Grizzle' bird.

A well-balanced Blue Chequer.

The distinctive 'Mealy-coloured' bird.

fanciers compare them to their own, especially if the loft has a good reputation or has housed a classic winner. It is always a joy to make these visits and enjoy the companionship and the theories of other enthusiasts. Many theories will be discussed, and one of those which I find most interesting is the wing theory. In particular one extremely learned fancier impressed me with his studies in this area. Simply put, it is the belief that the ability of a racing pigeon to perform best at either short or longer distances can be determined by the structure of its wing. This does not mean that any particular bird will be a winner if it has the proper wing type for a certain distance, but it can be classified as a potential winner in this category.

The wing of a long-distance racer, when spread out, will show very little enlargement between the secondary and primary flights. (You can use the diagram on page 58 in the chapter 'The young bird moult' as a guide.) The tips of the primaries will be more rounded than those of the short-distance flyer, which in contrast will show a pronounced 'step-up' or enlargement between secondaries and primaries, with the ends of the primaries more pointed. Good long-distance birds have a flexible wing which will open easily without the need for much force on your part when opening it, and the outside primaries will open up like the fingers of your hand, with spaces between the flights which are much broader than those of the short-distance exponent. It instead will usually seem to resist and snatch away its wing from your hand.

Of course there are other factors which affect the ability of a particular bird to do best over either short or long races, but the wing theory is one which has been tried and tested by many. In fact, I would not expect a bird of the 'long-distance wing' variety to do well at short distances, no matter how well trained, although doubtless they are capable of winning such events in the right circumstances and against the right opposition. However, I would stick my neck out and say it is almost impossible for a short-winged pigeon to do the reverse and win a demanding long-distance event. I said 'would' stick my neck out, but in fact I won't, because as you will discover in this sport, virtually nothing is impossible, and absolutely nothing can be taken for granted!

# TRAINING AND RACING YOUNG BIRDS

There is no magic formula to achieving success either with youngsters or with old birds, and I would warn all novices not to expect immediate wins. The novice who believes that by following my advice, or anyone else's, he will produce a winning team straight away, is under a false impression. No amount of advice will make a successful fancier, but what I hope to do is put the reader on the right lines and explain the methods I have practised with my own birds.

One thing which it is important to remember is that the successful fancier is the man who loves and feels for his birds, yet refuses to allow sentiment to change his system of management, so the novice will have to learn the hard way by his mistakes, and always remember that the man who never made mistakes, never made anything!

## Feeding

One aspect of the sport which has caused much controversy through the years, and continues to do so today, is feeding, and the fancier who formulates and devises the correct form of feeding is already half-way on the road to success. I feel that over-feeding is the downfall of racing pigeons, and with youngsters especially the amount of feed is vitally important. Birds do not grow with over-feeding. Instead they are inclined to mope and laze around when given too much – you probably know the feeling yourself! So never over-feed.

Although many magic formulas in the shape of pills and liquid additives are on the market, and claim to improve racing ability, records show that the speed and performances of the present birds are no better than they were 50 years ago when such things were unheard of. The late and legendary Belfast fancier, J. Kilpatrick, in his authorative book *The Thoroughbred Racing Pigeon*, could not, even with practical knowledge and endless laboratory experiments, arrive at a conclusive diet when he investigated this very complex subject. However, some statistics may be of interest to you concerning the actual protein value of the common foodstuffs.

Beans, such as tic, dun and soya beans, have between 20 and 30 per cent digestible protein, whereas cereal feeds like maize and wheat seldom have more than 10 per cent. What cereals do have is a much higher carbohydrate count, just under 70 per cent for varieties of maize, wheat and barley. Cereals have a low fibre content, often as little as 2 per cent, but an adequate amount of fibre, around 5 per cent, can be found in most types of bean, with the maple pea boasting a 5½ per cent count, and the dun bean nearly 8 per cent. Oil seeds, such as common sunflower seeds, are very high in fibre and although they should not be overdone, can be used occasionally to ensure that birds are not low in fibre. They are also high in digestible oils.

Many of the things that you will see pigeons pecking at when they get the chance, such as potatoes, cabbage and rice, have little or no protein value, but others do, particularly fish meal. Another high source of protein is yeast which is commonly used by fanciers to supplement diets. The idea is to formulate a steady, balanced, though not too complicated diet which the birds will thrive on. As for quantity, 1 ounce (28 g) per bird per day of an approved mixture, plus ¼ ounce (7 g) seed mixture, is sufficient for any racing pigeon when not feeding youngsters.

Happily, although striking a balanced diet may be difficult at first sight, the feed suppliers have taken the needs of birds into account in preparing their mixtures.

Where feeding is concerned, get out and about early in the morning. The birds love flying in the early morning air, setting off with a swing and appearing to really enjoy their work. After their exercise, which should be at least half an hour's duration, see that they trap quickly and feed approximately ½ ounce (14 g) of their corn mixture. In the early afternoon they can take ¼ ounce (7 g) of their seed mixture, either Haiths Red Band or Kilpatrick's Vigour Food. This can also be fed after a training toss with the balance of the corn left until after the last fly at night. To ensure that the correct quantities are given, weigh each day's feed. For instance, sixteen birds would get 1 pound (454 g) of mixture plus 4 ounces (112 g) of seed.

### Training birds to trap quickly

If a fancier is to win his share of young bird races, quick trapping is essential and therefore the youngsters should on all occasions be released for exercise on an empty crop. The same applies on training flights. If the birds trap quickly after each fly, you can be satisfied that you have played your part in their education. Accustom the birds early to the rattle of the corn tin – their signal to return to the loft immediately. If they have been on a long fly, encourage them to enter the loft immediately, not to wait on the roof.

### Your behaviour in the loft

It is in these early morning encounters with the birds that you can really get to know and enjoy them, just as they will learn to look forward to your visits with the corn tin. The manner in which you conduct yourself in the loft may not appear important, but it is to the pigeons. Watch any experienced fancier who has a good understanding of his birds, and observe that even on race days, he stays relaxed and collected, instilling calm and confidence in his racing kit.

Take your example from him and be your natural self around the loft. Keep the youngsters in while you are cleaning the loft as this helps them to get used to your presence. When calling in your birds, either by word or whistle, or rattling a tin, continue to do so when they are being fed. Frequently pick them up from the floor and handle them gently but firmly. Pigeons, like most birds, are timid by nature and you must gain their confidence and convince them that you are a friend not an enemy. Always make them at home inside the loft and never make loud or violent sounds as repeated behaviour of this type can have serious and permanent repercussions as far as a bird's confidence is concerned.

### Basket training

Once the youngsters have been flying around their loft for a few weeks, it is time to start basket training – educating them to eat and drink while in the hamper which will be used to take them to race starts. This should not be overlooked. I am convinced that many youngsters are lost in their first year because this very necessary part of their training has been neglected. This is especially true in the first and second young bird races when they are race marked and hampered on the evening before they are due to be liberated for their first fly home.

As a member of a racing organisation which sends about 20,000 pigeons for the opening young bird events, I am often appalled by the number of young pigeons that go missing, even from the most reputable lofts, and I have often inquired of the convoyers if the birds have been happy to feed and drink while in the crates. The reply usually is that few birds did. While many factors – such as insufficient training, clashing or birds being carried over their home lofts by others – contribute to losses, I am sure that the individual bird's chances of regaining its home loft would be so much better if it started off in good condition, having eaten and drunk all it needed, though never to excess. And this can never be the case if the bird is unhappy in the basket.

Educating your birds to be calm and eat and drink while in the hamper is essential if they are going to start races in the best possible condition.

If youngsters have been purchased from a reputable fancier, they will already be feeding and drinking, and usually they will also be accustomed to being in the hamper. If they are not, a few days and nights in a basket will be beneficial to them in later life. I am a great advocate of this basket training, and I have a trusty old thirty-bird hamper which I like to set on the lawn near the loft for this very reason.

I provide adequate food and water, put sand and wood shavings in the bottom and leave my youngsters inside. Be gentle and calm with the birds and don't overcrowd them. I then transfer the food and water – which for convenience can be put in jampots or anything similar – to the outside of the hamper as the birds are feeding, placing them where the young birds have to put their heads through the spaces between the wicker canes if they want to continue their meal. Adapting this method, I have even had the youngsters walk into an open hamper after being set on the lawn, so any fear they have of the hamper has been overcome once and for all. When catching young birds for their first training toss, they will not panic and scramble to get out through the crate or pannier, through which they might be injured, if this fear has been conquered.

Good basket training has paid off in this case. The bird is obviously contented as it awaits transportation to a race point.

### Road training

When the youngsters are flying well around the loft, the next step is actual road training, and this is another issue upon which fanciers often disagree. Some believe that young birds should be trained over short distances, preferably on the line of flight of their first race. All that is necessary to start them off is a short flight of 1 or 2 miles (1.6 or 3 km), increasing this by stages of 15 to 20 miles (24 to 32 km) with the ultimate target of training them systematically towards the first race point, or as near to it as is possible.

Always choose good weather conditions for your training tosses. For their first flight, all that is needed is a short flight of one or two miles, increasing this by stages of fifteen to twenty miles.

Some fanciers even have the birds at this stage before official racing commences. However, I favour the method of having at least four tosses in different directions, releasing the birds north, south, east and west of the race location. When this has been completed I continue on the line of flight, training up to a distance of 50 miles (80 km). My club's opening young bird race is usually just over 100 miles (160 km) in distance and I have found that the resulting increase of distance of 50 miles (80 km) suits my management.

### Training tosses

How you take your birds to the liberation point for the first toss depends, of course, on the facilities available to you. If you are not fortunate enough to have your own suitable form of transportation, you will need to speak kindly to a fellow fancier who would be prepared to take your birds along with him.

Evenings are the best time to liberate the birds, but whatever time you choose, pay particular attention to the weather conditions. It is better to bring the birds back, never having left the basket, than to liberate them in unfavourable conditions when likely as not you will be liable to suffer the great disappointment of losing several birds, if not the entire kit. Avoid mist or fog, rain or storms and cold easterly winds.

I have already mentioned the importance of ensuring that the birds will take food and water in the basket when being taken to the release point, but when training always basket birds on a light feed, and allow them to settle for a time at the liberation point before opening the basket door. A tip here is to send a reliable and experienced older bird, if you have one, for the first few tosses as it will have a steadying influence.

If accompanying the birds for their first toss, liberate them together, watch them carefully, and take note of their line of flight. If they show any tendency to fly around or swing away in the wrong direction, and are an unreasonable time in making for home, it is advisable to give them at least one more toss from the same direction or point before continuing with their training schedule. If this occurs continually you must exercise patience and apply the same principle, at least until the birds are heading for home immediately. Be kind enough to offer them a small titbit on their return – they will have deserved it!

Observe all your birds on training tosses so that you can assess which ones to put in for forthcoming races.

If carried out methodically, a few tosses should ultimately see them head for home as soon as they are released. At the half-way stage of their training, if the birds are well-steadied, they can then be released in pairs, rather than all together, and as training progresses, single liberations can be included so that the birds have to rely on themselves and cannot follow other birds. Once you know they are not 'following-the-leader' you can revert to the normal training routine.

I must stress that before the opening race, regular training and patience on your part are of considerable importance. Young birds cannot be trained too much, but of course they can be over-raced. Always remember that the youngsters in their under-developed state can show obvious signs of fatigue if raced hard in a series of difficult events, and this can in fact prevent them from maturing into good healthy old birds. Regardless of this, as a firm believer in the survivial of the fittest, any that show a kink or defect of any description should be disposed of immediately.

During tosses you can watch and compare the individual performances, choosing the outstanding birds for earlier races. Refrain from training them on either of the two days immediately preceeding a race unless the weather conditions are ideal, and even then a maximum 10 or 15 mile (16 or 24 km) spin is all that is needed to sharpen them up.

**Racing in the first season**

After beginning his stock of birds, one of the most important problems faced by the young fancier while still in the process of building up a team is how far he should race his pigeons in their first season. Opinion is often sharply divided on this matter. Some believe that young birds do not suffer by racing the complete programme, while others think it best to give youngsters just one or two races for future education, rather than subjecting them to a rigorous first season. Much depends on the individual fancier's plans and ambitions for the future. If the aim is to concentrate on normal club racing up to 200 miles (321 km) where the emphasis is obviously on speed, it is likely that the all-out programme has its advantages.

For the young fancier building up stock for long-distance racing, caution is necessary, and his chances of succeeding in this respect will improve if he limits the amount of club racing he expects his youngsters to do in the first season. At the same time, if he has the numbers, he may be tempted to try one or two in the National or Classic events, where the distance is usually about 300 miles (482 km). There is no reason, provided he can afford one or two inevitable losses, why the novice should not compete against his more experienced rivals in this calibre of competition. Provided they are fit and sound, young birds if flying well require less of the owner's time and attention than old birds, as they have fewer domestic chores such as laying the eggs, and feeding and rearing the young birds in their nests. Success really depends on training and fitness, and therefore they require a methodical education before being sent for their first races.

How far you decide to race your birds will ultimately be a personal decision, but as a guide I believe that young birds receiving a thorough basic training can be raced up to 250 miles (400 km) without suffering any after-effects. Much does depend, however, on the route and the type of race. An easy inland race of that distance is an entirely different thing from an event of similar distance but where the young birds have a cross-Channel or water-crossing section, to negotiate, so if you do decide to race your young birds, make sure you choose the races carefully.

## Racing young birds

Young birds are usually raced to the loft and the feeding dish – in other words they come back because they are hungry and they know they will be fed in the home loft.

Although all sizes, shades and colours of pigeon win races, I personally prefer the medium or smaller bird, well-balanced and able-bodied, choosing the first round youngsters for the first and shorter races, and the second-nest birds for the longer events. My reasoning is that the later-reared birds are not so advanced in the moult, the wing moult as well as the body moult requiring consideration. This is discussed in detail in the chapter on 'The young bird moult' on pages 57-60.

Generally, the main considerations are speed and the ability to trap quickly. Usually only a few yards per minute separate the leading birds in a short event. Even if your young bird comes racing home, alights like a bullet on the loft and traps eagerly, to provide you with your first success, don't get over optimistic and imagine that you have a future champion on your hands. I have had in my own loft, young birds that were seldom out of the clock as youngsters yet never fulfilled their promise, showing nothing like the same ability when their first year was behind them. Many other examples could be given of young birds being backward in their first year, but turning out to be prolific winners as old birds, and more still of youngsters not raced in their first season and making their first impact on the racing scene as yearlings.

As far as the beginner is concerned, once he shows ambitions of flying the longer distance events, neither success nor failure with his first season's racing should be accepted as an indication of how his birds will perform in future years, and therefore a sensible programme of young bird training must initially be of the utmost concern.

When racing commences in earnest, give as much attention to the late arrivals as you would to the earlier birds and always remember that the latecomer one week could well be the first pigeon home from the next race. Avoid clocking more birds than necessary as this may result in bad trappers.

Whatever number of birds you train, and whether they are old or young birds, I would divide them into two distinct teams, sending them in alternate races no matter how strong the temptation is to send them all. That way, if you are unfortunate enough to encounter a difficult race when losses are high, you will have fresh birds for the following week. All too often a fancier has seen his first stock of young birds lost in a disastrous race and then has to face added misery because he has none left to race for the remainder of the season. There can be fewer sports where the uncertainties are greater, for it takes just one bad weather forecast, just one little freak occurrence among the elements, to trigger off a mass loss. With all the technology that goes into the making of weather forecasts nowadays, these should not happen, but they always have done and always will, and when they do, the last thing you want is a situation where your racing stock for the season has been decimated at one swoop.

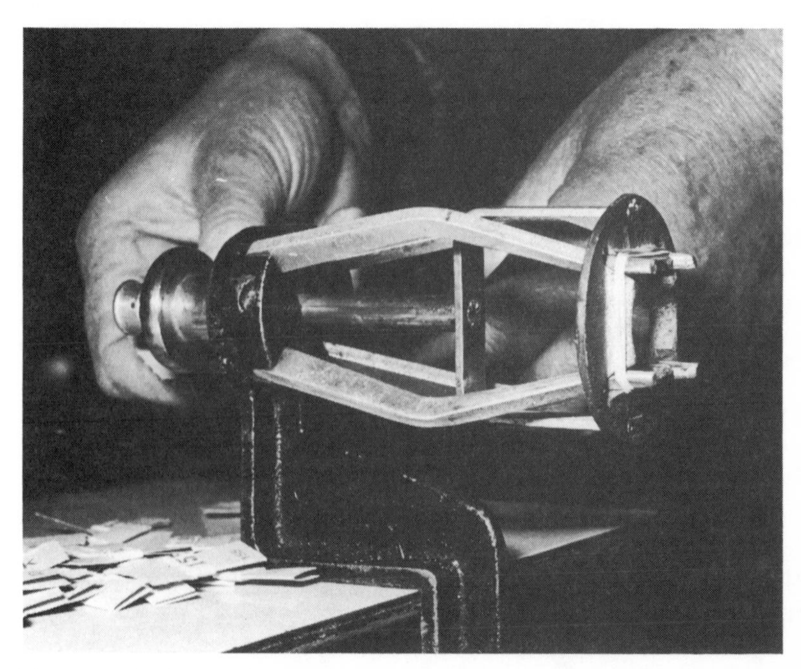

The race ringing device used to put the rubber ring on the bird's leg.

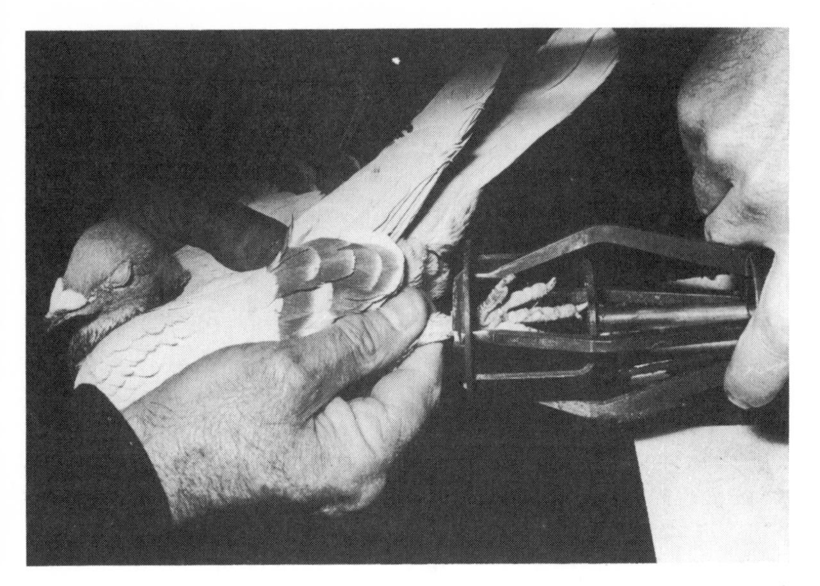

It may look painful, but nothing could be further from the truth – and old birds like this one have seen it all before.

Young bird racing is regarded primarily as a necessary preparation for the several seasons of old bird racing that lie ahead if the youngsters are not abused. Therefore not everyone will be interested in this aspect of the sport, and indeed many well-known long distance racers would never contemplate racing their younger birds, believing that successful youngsters respond mainly to the corn tin. This means keeping them peckish and under-fed to encourage them to fly back to the loft quickly, and therefore their future growth and development may be adversely affected.

It certainly makes a difference, for the experienced fancier will quickly notice the contrast when he handles and compares the tightness of feather, growth and development of young pigeons that have been overworked, compared to those that were given an easier time.

To me, young bird racing is fun, and would be lost without it during the months of July, August and September. However, I am all in favour of arriving at a suitable balance. Like others in the sport, I delight in any winnings or successes I can derive from the youngsters, but I always try to save enough of them to compensate for the losses which are inevitable when the old bird team races to very long distances.

### Pairing up young birds

Over the years, the tendency to rear early youngsters and have them paired up for longer races has increased, and many well-known fanciers are attaining a fair measure of success with the following management.

When a couple of youngsters are seen pairing up in the loft, don't separate them, but make the best use of this condition. Place a nest pan in the corner of the loft floor, or in a nest box when they have taken to it, and to prevent the young hen from laying, place a couple of dummy eggs in the nest, one at a time. This will usually satisfy the pair's natural instincts and they should continue to sit indefinitely. On basketing day for the race, slip a small youngster into the nest for a few hours before hampering for the races, and make sure that both have a turn at sitting on the youngster. When they return from the race, don't put the youngster back in the nest – let them continue sitting on the eggs.

If, in the process of building up your team, you have acquired some old pigeons and by chance one of the old cocks takes a fancy to a young hen, and she responds by following him into the old bird section of the loft, again take advantage of the natural situation which has arisen, although take care not to leave them together during the night – an hour or two during the day is ample. On basketing day keep them apart until it is almost time to hamper your entries. Then slip the young hen into the old bird section, having previously given him another hen in his nesting box. Allow the young hen to see him through the nest box front for five or ten minutes, then catch and put her in the basket.

Common jealousy will make this young hen want to get back to the loft and her selected mate as quickly as possible, so when you are pooling your birds for the race, don't leave this one out!

These are two obvious examples of how you can take advantage of your birds' behaviour to get the best out of them on race days, but always keep a sharp lookout for any little fads or fancies, and make the best of them in the same way. This is where the keen and observant fancier comes into his own, treating each bird as an individual and ensuring that a particular pigeon's habit works for him, rather than against him.

### Fly-aways

As already mentioned, many fanciers, old hands and novices alike, have at one time or another experienced the unfortunate and dreaded 'fly-away' where a young bird suddenly takes to the air, never to be seen again. Usually the helpless fancier is left asking himself how and why it happened, and how he can prevent further catastrophes. From information obtained from victims of this unhappy circumstance, these losses, which are a yearly hazard, sometimes of colossal proportions, usually occur when young birds are from 11 to 14 weeks old, this being the period when their squeaker eye is changing to the natural eye colour of the adult pigeon.

Another important factor seems to be that most fly-aways take place during early morning exercise. This is something of a phenomenon among fanciers and the only advice I can offer consequently is to be extremely careful with the youngsters while the eye-changing process is taking place. Don't, under any circumstances, give them access to the open trap at this time of their development – otherwise you may well lose the lot.

### Late-breds

Many fanciers have a tendency to rear a few late-bred birds, usually bred from their best old pigeons at the latter end of the season in late summer, and in most instances they are hatched from birds that have been engaged in long-distance events. The question often arises in the fancier's mind – would it be better to train them a few miles, or let them grow and develop, resisting the temptation to educate them by training until the following year?

To train late-breds late in the season is a dangerous practice, as they will inevitably have started moulting in an irregular fashion, sometimes throwing two flights at a time instead of one as they would normally do if reared at the proper time of year. Another disadvantage would be that the body moult would also have started, which reduces the bird's chances considerably if the weather turns bad.

As these late-breds are derived from the best pigeons in the loft, I would advise that they are not trained until the following April, May or even June, when the feathering, which after all gives the bird its ability to fly fast and accurately, is in much better condition. Although there will always be exceptions to this rule, the percentage of good pigeons reared as late-breds is very small, the main reason being that they are bred out of natural season and are therefore much more likely to suffer a set-back in their health in the colder autumn weather instead of having a normal moult when the weather is warmer. By the time the late ones are moulting, the early winter is often upon them and their natural protection against the cold is gone.

# THE YOUNG BIRD MOULT

The transformation in the appearance of a young pigeon during the moulting process, when they change from the dull-feathered, mauve-eyed youngster, into the adult pigeon, has always astonished me, and may even cause concern to the inexperienced fancier. It occurs in the summer, and although gradual in the early stages, it reaches a point when the youngster is almost naked, bare of feathers, with the neck and head appearing as if they had been plucked. The tail and wing moult are also of importance to young bird racers, for it is essential that they know the order in which the tail and wing flights fall.

The moult, or casting of the feathers, begins when the youngsters are approximately six weeks old, with the last remaining tail feathers held till the young bird is about six months old. Novices may initially get the impression that the moult is some horrible disease, but nothing could be further from the truth, for it is a simple and natural function. Nature decrees that most animals and birds shall in part, or entirely, renew their fur or feathers annually. If the procedure did not take place, the feathered species would in time be denuded of feathers and would of course be unable to fly.

Because of the orderly process of the moult, it imposes the least possible trouble and inconvenience when the bird is in flight. Some families of pigeons cast their feathering sooner than others. To help with the moult, birds should be provided with frequent baths, plenty of clean water to drink, variety in feed, and sensible management.

However, pigeons are more susceptible than usual to ailments during moulting. It is a very testing and difficult period, draining their energy system because of the need at an early age to sustain and supply vital blood and sap to the growing feathers. However, if they have been well nourished, and not worked too hard in racing, their health need not be adversely affected in the slightest. A good moult is in fact a sure indication of good health and proper management, and you can take it that a good first moult augurs well for their present health and future well-being.

If the opposite happens, and there is a certain sign of weakness with the bird in poor condition, it is unlikely that the moult will be satisfactory. Young birds hatched in the normal breeding season usually have the least difficulties in moulting, while the young hatched after July and termed as 'late-breds' only effect a partial moult, with some of the primary and secondary flights being retained until the summer of the following year. I have noticed also that youngsters bred very early in the year also achieve only a partial moult, losing and growing new feathers on the head and neck, only to start all over again when the general dropping and renewal takes place.

The wing comprises twenty-two feathers, with the ten outside longer ones called the primary flights, and the twelve smaller ones called the secondary flights. On rare occasions I have seen birds with eleven secondaries and the same number of primaries. The first feather to fall is the short primary flight in the centre of the wing. It should be cast without any difficulty by a strong and virile pigeon. The new feather will slowly but steadily develop and take its place, firstly in the form of a small bulb protruding from the follicle, then, as it gradually grows, it uncovers the webbing and a new and strong feather takes the place of the one lost.

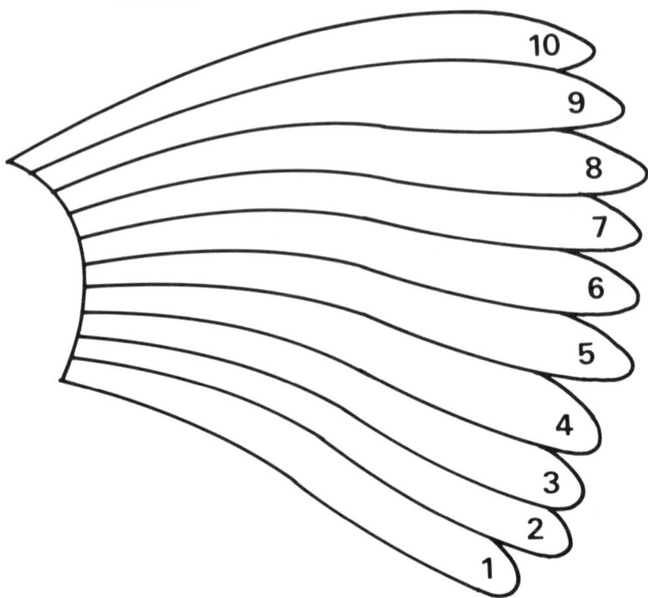


This shows the sequence in which the ten primary flight feathers of the wing are cast during the moult.

In about four weeks' time the second primary is dropped and when this is half grown, the third falls, and so the sequence continues, the remaining flights following at longer or shorter intervals until the tenth and last is dropped. There is no rigid pattern of moulting for the secondaries. The interval between moulting the flights and the secondaries can be considerable. By the time the fifth primary is dropped the moult progresses more rapidly, while continuing in a methodical pattern. The small feathers at the base of the back are the first to fall from the body, followed by the entire head. The neck and breast are next to be stripped, then the feathers on the rump or back, and so on to the tail.

The moulting of the tail is a most interesting process, as the tail's purpose is two-fold. Firstly, it acts as a guide, or steering rudder, and secondly it forms an inclined plane to support the posterior of the bird when in flight. As it also assists the bird when taking off and landing, the tail is obviously of great importance, as is the moult.

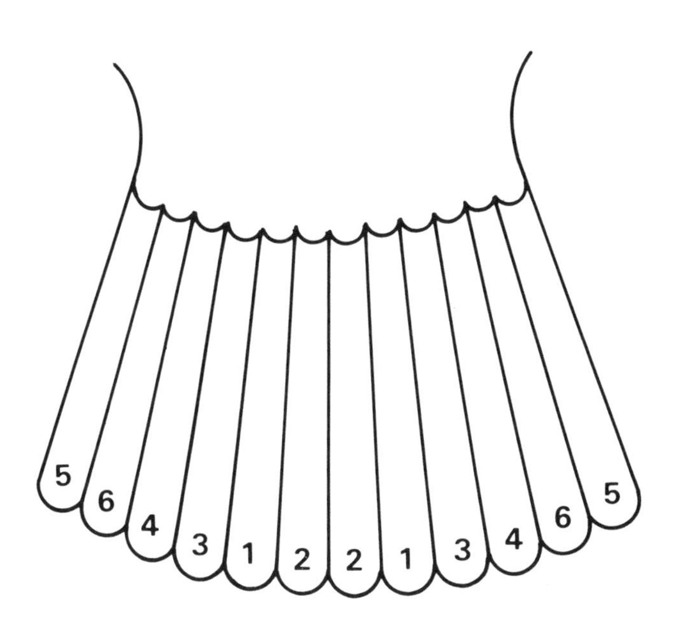


This shows the sequence in which the young bird's tail feathers are cast during the moult.

The tail has twelve feathers called rectrices, six on each side, with the two inner feathers called medians. The first to fall or come away are – contrary to what many people believe – not the centre pair, but the pair on either side of these medians. These are generally followed by the medians, then the remaining feathers from the centre towards the outside of the tail, although there is no absolutely rigid pattern of moulting for the tail flights.

The pigeon's feathers are renewed each year as the bird gradually loses all its old feathers and replaces them with new ones. The process is always gradual, starting with the feathers of the head and ending with the tail and flight feathers, in order to cause as little inconvenience as possible to the pigeon during the course of the moult.

# SHOWING FOR PLEASURE

Showing pigeons is rapidly growing in popularity and nowadays the most dedicated of racing enthusiasts is finding pleasure through participating in the winter shows staged by clubs all over the country. The advantage of weekly shows is enormous as they benefit both the club and its members, helping to maintain incentives at a time when racing is out of the question, and offering another avenue for success. If you have had a bad racing season, never mind – the shows will be there in the winter months ahead to give you something else to aim for. I think it also benefits the birds, for shows mean that they cannot be neglected in the winter months. Even at the club level, these events also create a wonderful social atmosphere and a forum for discussion when fanciers from neighbouring clubs are invited to adjudicate.

The author (right) with a happy owner after I had awarded his fine birds a first place in a club show. In the background, you can see the show pens containing some of the other birds.

The club series is usually completed by the end of the year and the revenue is important to the clubs, even though the cost of entry may be only minimal. Many clubs then allocate the profits as Show Special cash prizes to add to the interest.

Although most clubs will be honoured to accept entries from non-racing fanciers, termed professional showmen, these events depend mainly on the support they receive from the racing lofts.

Controversy among the fanciers, good-natured though it may be, is never far away, for beauty is in the eye of the beholder and therefore judges will inevitably disagree. This fact has to be sportingly accepted by all fanciers who enter shows in which, contrary to racing, it is the man and not the clock which judges success. Nevertheless, anyone accepting the arduous and sometimes unpleasant job of judging should make it a priority to vet and handle each bird individually and be able to give a detailed judgement of his own personal opinion, discussing the pigeon's weaknesses and strengths.

Often the birds will be judged 'through wires', and in this case the birds are not to be handled. The judge will make his selection on what he sees, and the club responsible for organising the show should ensure that none are removed from the pen for the judge's inspection – at least until the cards are awarded.

My own first experience of judging many years ago was in the company of a team of well-known fanciers with the task of judging a 'flown class', and a disappointing inauguration it turned out to be. With all competitors out of the room, as is usual, the judges walked along the pens and after noting several numbers decided only to handle a few selected birds, basing their judgement on this. My conclusion was that it was a most unsatisfactory and unfair way of arriving at a decision, for all competitors pay their entry fees and it is only natural that all should be vetted to the same extent. In all probability the best bird was never out of its pen. Unfortunately there is no set of rules by which a judge must arrive at his decision but I believe that anyone with this responsibility should be 'fair to a fault' if that is what is needed to please the fanciers.

One of my own favourite show classes is for mated pairs, when very often fanciers have on show the birds they intend to pair together for the coming year. The opinion of the judges will often be valued by the fancier wondering if he has put the right birds together. As far as the show entrant is concerned, if you

If your birds are accustomed to being handled in the loft, they will take much more kindly to being examined by the show judge.

are fortunate enough to be blessed with a proven breeding pair, show them by all means and test the knowledge of the judge. Bear in mind that the racing pigeon in strange surroundings will not always 'show' as well as it might, and I have even seen penned birds doing their utmost to free themselves through the wires, while others will lie in the cage, hiding their charms. Therefore it is advisable that fanciers should have something resembling a show pen or nest box with an open front which could be used to train the birds for showing. It will have more confidence when penned at the show and will consequently give a better account of itself. Place the practice show pen somewhere a bit noisy where there are people moving about, in the living-room if you like, as shows are usually fairly bustling affairs.

Pigeons with fret marks, crooked keels or in generally poor condition, do not appeal to any judge and it is pointless to show them on the off-chance that they will be selected for a prize, so disregard them when choosing your entrants. Make sure the bird you choose is clean, particularly its feet, and is in the pink of condition. A little extra care before a show can make all the difference.

One more point of considerable importance to the novice showman is the necessity of making sure that the birds are trained sufficiently to enable quiet and easy handling by the judges. Some judges will use a judging stick to gently prod the bird and make it posture. Try to make your bird used to this by gently nudging it with a similar stick while it is training in its show pen.

# MANAGEMENT AND RACING OF OLD BIRDS

By now you will be familiar with the young birds, and may also have taken an interest in showing as well. The time has come for contemplating the racing of the first year's birds as yearlings in your local club's old bird programme. The great and unquestionable secret of successful old bird racing, as with the young birds, is their condition, and although there are doubtless exceptions to the rule, I have never known an eminent fancier who uses 'magical' chemical-based formulas on good pigeons in the pink of their condition. Putting it bluntly, my advice is to turn a blind eye – or at the very least a sceptical one – to the much-vaunted tonics and medicines which, it is claimed, will make winners out of mediocre birds. Instead, put your faith, confidence and energy into producing birds in tip-top condition for races. To do this, you will need to apply correct, methodical management not just for the few weeks leading up to the racing season, but for twelve months of every year. When that condition is achieved, rest assured that the remainder, in a sense, is a question of the bird's own racing ability.

In many ways the same applies to pigeon racing as to horse racing and greyhounds. Even the best breeding progeny can produce poor class performances if the candidates are not in form, and they can even be outclassed by less well-bred stock on the day if the poor relations are sent off in better condition, providing that all other considerations are equal.

**Treating birds as individuals**

Many other aspects of pigeon racing present endless problems. Apart from condition, the individualities of the birds is a subject which will give the novice much food for thought. No two birds appear to produce their best racing form under the same conditions. One may race better to youngsters in the nest, another will respond more eagerly to eggs. The temperamental peculiarities of birds require a great deal of study and thought, and if they are ignored there will inevitably be the disappointment of poor performances with the fancier unable to get the best out of his stock.

You can also expect difficulties with birds which you have not bred yourself, but which have been introduced into your loft. Although I cannot say why, it has usually been my experience that the majority of these birds, however well trained and cared for, and regardless of how well they have been prepared, seldom do as well as home-bred stock. In many instances they will be lost, not because they are inferior, but because they appear to have less loyalty to the loft.

My theory – and it is only a theory – is that birds are ill at ease if brought to race to a new loft in their first season, and that they require extreme care until they are in their second or even third season before being asked to perform from the longer distance events. As in all aspects of this complex sport there will be cases to prove an opposite point of view, but I believe that, if the truth were known, the majority of 'acquired' birds are lost. However, young pigeons which voluntarily take up residence at a new loft will often race better to it than to their former home. I know of

An Old Bird Derby winner. This white-flighted Blue Cheq hen won the INFC Old Bird National from Nantes in France for A. Simpson and Son in 1974.

instances where this has happened, and some forlorn fancier who has agreed to a transfer is left wondering why the unexceptional youngster he reared has suddenly turned into a winner for someone else.

### Competitive racing

The competitive season usually commences with mating, which in my early days was from February, but now happens earlier in the year. Now it is quite common for fanciers to have youngsters in the nest ready for ringing when the rings are released by the unions early in January.

At present there are two systems widely used for the racing of old birds. The most popular is what is termed the 'natural system', and the other is described as the 'widowhood system'. I have found that the former has worked better for me.

Training should start soon after the birds have hatched and reared their first round of youngsters, and are sitting on their second round of eggs. When nicely settled down, and when weather conditions are favourable, choose a fine day for the first toss and try to avoid training when there is an easterly wind. Training should start at approximately 15 or 20 miles (24 or 32 km) gradually increasing the distance until they are close to the first race distance. If possible I give mine at least a couple of tosses from the actual first race point. Provided the birds have been exercising well around the loft, there is no need to be as thorough as when they were youngsters. Some fanciers, and successful ones at that, use the open trap system in their loft where birds can go out and in as they please, but this is not suitable for everyone. If it suits you better, give the birds a fly at a set time each morning and evening.

In most circumstances the hen takes over the incubation of the eggs during the late evening and will continue sitting through the night, with the cock bird taking over for the daylight hours, so it is reasonable to exercise them before they take over their daylight shift. I put them out very early in the morning with the hens taking their exercise in the evenings. If the birds do not seem too keen on flying, or indeed if they refuse to fly for at least an hour, a makeshift device such as a flag, waved when they settle, should keep them on the wing for the appropriate time before they are allowed to trap for an evening feed.

The author (right) and a keen fancier admire a fine pair of birds.

**Racing yearlings**

In the natural system birds are raced when sitting on or feeding youngsters and the young fancier will be faced with the dilemma of competing with his team of yearlings against the established lofts in his district, where the owners have more experienced birds of varying ages. While he is at a disadvantage in this respect, there is no reason why he should not achieve a fair measure of success if his stock are of good quality, provided of course that his attention was not lacking during the off-season.

The distance to which yearlings should be raced has given rise to much argument and controversy with many successful fanciers sharing the opinion that yearlings should be treated with the same caution and care as was accorded in its first season, and that 250 or 300 miles (400 or 480 km) is the maximum that can be expected of them.

By and large I agree with them, but it has to be admitted that outstanding performances have been accomplished by yearling birds over distances well in excess of 300 miles (480 km). The 1980 Old Bird National winner from Les Sables in France to Ireland – a punishing flight – was a yearling, flying 583 miles (938 km) over what is claimed, and rightly so, to be the world's most arduous and difficult route. Once again the deciding factor will ultimately be determined by the particular circumstances, and the aspirations the novice holds, but I would advise a certain degree of caution in the management of a yearling team. There is no advantage in risking heavy losses which would lower the standard of the breeding team which was intended to form the nucleus of a successful old bird family.

A strong team of yearlings is the backbone of a successful loft, especially where the novice is concerned, as these birds will be competing for him in the years to come. I would urge that at the most, only half of the yearlings should be sent to the 300 mile (480 km) stage, the rest being retained to maintain stock and replace losses. Once the birds have accomplished their first flight of this distance they can, with more confidence, be engaged in the longer distance and classic events in future years.

### Choosing birds for races

Another of the advantages enjoyed by the experienced fancier over the novice, especially in old bird racing, is that he knows his birds well and finds it easier to prepare them and have them at their peak at just the right time. Some fanciers prefer hens for long-distance racing, believing that when caught at their peak, they are more dogged and determined in their attempts to reach home than are birds of the opposite sex, but I have always found them to be much more difficult to prepare and get into top condition, mainly due to their tendency to get 'eggy' at awkward times.

Care and caution should be exercised when deciding to send a bird off for a race. On no account should a hen be basketed when about to lay. This is an easy circumstance to diagnose if she is about to lay for the second time, but you may have trouble over the first egg. Some hens take more time to lay up than others, but usually, the first egg will appear five or six days after the cock is driving hard. Take the precaution of feeling the vent bones before basketing, as shortly before the egg is laid these will start to move apart – nature's way of allowing the egg to pass unharmed between them. If you are familiar with the feel of the bird then this movement is very noticeable – another advantage of getting to know birds, particularly hens, as individuals.

Like any other livestock, pigeons vary greatly in their reactions to changing circumstances and to a certain extent the performance of a bird when racing will depend on the emotional changes it goes through when mating. One bird may reach its best form when sitting on eggs for three or four days, while others may 'peak' when the eggs are about to hatch. Taking all things into consideration, hens race well when on chipping eggs, or to a newly-hatched youngster; while for cocks, chipping eggs, or a ten- to fourteen-day-old youngster, and calling his hen to nest again, are also favourable conditions. As well as this the fancier must look very carefully at the likes and dislikes of his individuals if he wishes to be successful.

This is when you can indulge your birds a little when it comes to feeding. A bit of extra, hard flesh will not go amiss over the long races, but you must not mistake fat for muscle. When you hold the bird, if it feels heavy and flabby, then it is too fat. A bird of just the right weight will feel hard and 'corky', with soft and

A fine bird perching on top of his nesting box.

silky body feathers. Hens in good condition will usually sit on the nest box perch with tail quivering, eyes alert and blinking, and paying a good deal of attention to the nest while the cock is sitting.

'Cocks driving' is a condition which, like all others, can produce either success or failure. Yearling cocks, sent to races while driving their hens to lay, have romped home well ahead of their loft mates and won many a good race, but more regularly, when sent in this condition, they fail to reach home at all. At this particular stage yearlings are extremely excitable and should not be put out to race until their hens have laid up. Older and more experienced cocks often race well when driving the hen to lay but so much depends on the temperament of the bird, and it is always a risky business. I believe firmly that each cock and hen has a parental incentive that encourages its racing capabilities to varying extremes at different stages of the sitting process, and finding that exact time needs careful study on your part. If you combine good conditioning with a strong parental incentive, you have a factor essential to successful racing. This parental incentive is very noticeable in yearlings, some of whom improve while others become worried and over-excited, causing a deterioration of their physical condition which will in turn affect racing performance.

The effects can be very misleading in a different way. I have discovered that, with my own racing yearlings, whether cocks or hens, the best time to expect results is about three weeks after the fall of the first flight so that the birds for these events are mated accordingly. Long distance racing, that is, events of over 300 miles (480 km), when every extra mile counts, is really the supreme test of your family, and birds which keep your name among the prize-winners on a consistent basis are in a class of their own. These occasions demand extreme fitness, with birds on the wing for long hours, without water, food or rest. Therefore the management of these long-distance flyers is entirely different from the system for shorter inland races. These longer races are flown in late June or July, and most organisations decide upon their programmes during the winter months. Once the race dates and points have been finalised, you can then pick out the birds you will want to enter when summer comes, and if you have birds which have flown across water before, so much the better as they will not require so many short tosses.

### Mating of selected birds

The mating of these select birds should not commence until early April as you will want to prevent them from moulting. This will stop them casting their first flight, which will fall after the pair have been on their second round of eggs for about ten days. It will not always be so but if they are in good condition this will usually happen. With both birds nicely down on their eggs, and conscious of building up their energy for the big event, the question arises whether to let them feed or rear any young from the first nest. My advice would be to let them hatch and feed the youngsters until they have reached ten or twelve days old, when they could be transferred to another pair feeding youngsters of approximately the same age, to allow your racers to be brought slowly and methodically to the peak of fitness. If racing them before a big event, choose the race carefully, possibly including one cross-Channel race in their preparation at least three or four weeks before the 'big one'. After this preparatory race, and as soon as the birds are sitting again, you can recommence their daily exercise and if they do not fly to your satisfaction, revert to 40 or 50 mile (64 or 80 km) tosses twice a week. On these days allow no further exercise. If possible, give them access to the 'open door' and the ideal condition would have them sitting ten to fourteen days on their second round of eggs on race-marking night. Most birds, especially hens, are very keen on a small youngster, and it can help to slip one to each pair on the night before marking – a little ploy that has often brought birds home in winning time.

### Widowhood racing

This method has been practised successfully, particularly by Continental fanciers. The legendary Fred Shaw, described as the Gurnay specialist, had phenomenal success in the Manchester area with widowhood racing, and no doubt his exploits have inspired many modern fanciers, for the system has become a craze, taking the fancy by storm.

I should point out that it is not a system I have keenly practised myself, but many fanciers known to me have used it well. From my own observations, one drawback worth considering is that for racing purposes, none of the hens can be engaged,

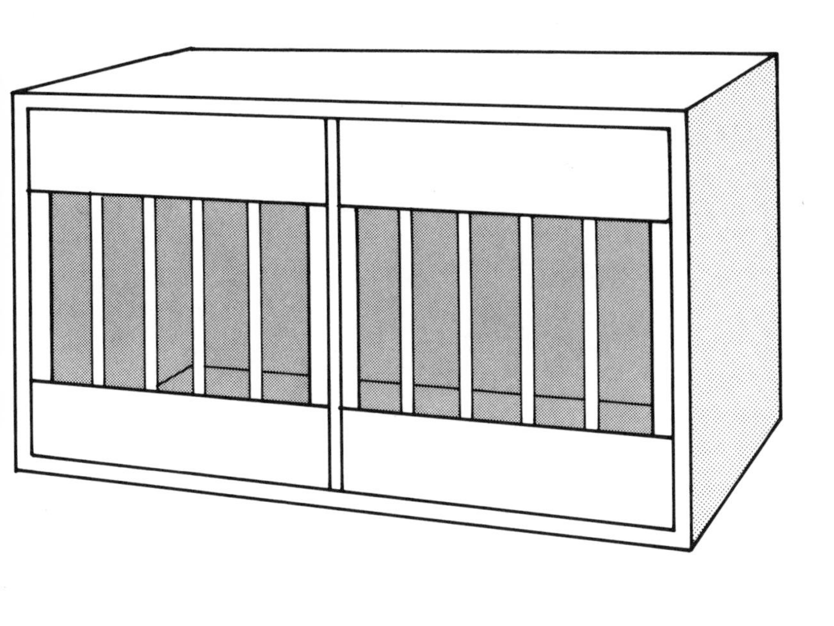

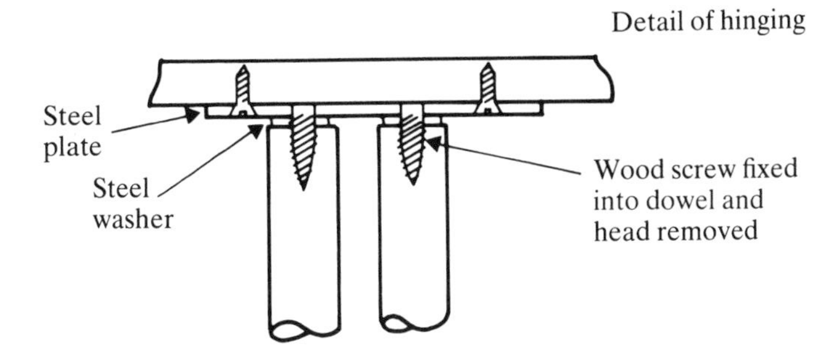


A widowhood nesting box.

and quite possibly a potential winner could be left sitting in the loft, acting as nothing more than an incentive for its partner.

According to the Belgian fanciers, nest boxes for the widowhood system should comprise of two compartments, as near as possible of equal dimensions, with the length just double the depth (see diagram page 73). The idea is that one of the nest box halves, which are hinged in the centre, can be folded back, allowing the hen to be fastened up without the cock being able to contact her. Through this arrangement they are able to bill, coo and carry on their courtship to some degree. Therefore when the cock is sent racing, he will be all the more eager to make a swift return to continue this relationship.

A difficulty is often encountered in achieving satisfactory moulting, so the pairs should be allowed to sit only for six to ten days before the hen and eggs are taken away and in the case of a cock which has had a backward moult the previous season, he should be allowed to rear a single youngster to between ten and twelve days old. When the hen is taken away the loft should be kept quiet, the cocks not allowed even to fly around it, but fastened up in their widowhood box where they should also be fed and watered. They should not, of course, be prevented from taking their daily exercise.

Once the hens have been removed out of sight and sound from the loft, they too should be kept in a separate box to prevent them from mating up to each other and laying. The hens should be moved back to their original nest boxes once or twice a week, ready and waiting for the return of their partners from a training toss or exercise. When they return the cocks should be put in the widowhood box with hens so that they cannot reach them. After about an hour, the hens should be taken away and the loft, perhaps through the use of blinds or shutters, should be left in semi-darkness.

You must be prepared to sacrific some corn in the use of this system as the biggest headache is persuading the cocks to feed. If you know that a bird relishes a particular grain, ensure that there is plenty of it in the mixture to tempt him.

The cocks should be exercised or tossed three times a day, preferably as early in the morning as possible and as soon as they alight at the loft they must be encouraged to enter their nest boxes for feeding. For training, short tosses are more satisfactory than long jumps. When he returns, with the hen waiting, make sure that he does not 'tread' her – take him away when she becomes too keen. Choose fine days for training, and after the race the cock should be allowed access to the hen for a couple of hours. Also treat them to a bath and a feed of whatever they fancy. Then the hen should be put away again until the middle of the week when the cock can see her again following his return from a training toss.

Handle the birds often to check that they are holding their weight, and take particular care with the hygiene of their nest boxes. Always carry a small cane or rod into the loft and if the birds are becoming too noisy, draw it across the dowels on the nest box fronts to quieten them – they will get used to this little discipline. Allow them on the floor only for a few minutes to take a spot of seed before exercising. Leave them to themselves as much as possible and be careful that the nest box fronts are carefully secured because if a bird gets into the wrong nest box, it will be sheer murder! After the morning exercise, feed them with seed mixture, at mid-day give a light corn mixture and in the evening feed with tics, taires and maize.

The cocks should be exercised three times a day when racing on the widowhood system.

The widowhood system has its variations, the one I have described requiring two lofts for the successful separation of the birds, but I know cases where the cocks and hens can be raced to one loft.

One of the reasons why the system is so popular is that pigeons flying 'widowhood' are quieter and calmer in the basket as they are accustomed to being fastened up in the loft nest box, and rarely do themselves any physical harm. Birds raced on the natural system, unless well trained in the basket as youngsters, do not settle as well, and if basketed for racing when feeding a well-grown youngster, will fill themselves up with corn. The chances of this bird doing well are almost nil and of course there is the added risk that the corn will turn sour.

## Jealousy system

For those with time, patience and a desire to experiment, I would recommend this system. It could be construed as another form of 'widowhood' and could be indulged in with one or two early bred young cocks or yearlings. It requires a good, roomy nest box divided into three compartments, preferably with a dowelled partition for good vision. Choose two young cocks of the same colour and pair up to an old hen. Then, putting the hen in the central compartment with one cock on either side, allow them to vie for her favours. The hen will do plenty of bowing, going from one cock to the other and making each of them jealous with her affections towards the rival.

Now and then the sliding partition can be taken out to allow one of the cocks to keep company with the hen for a few minutes, but keep a watchful eye to make sure that no treading takes place. Then replace the partition, giving the other cock his chance, and so on, while training and exercising continues. On basketing day the birds should be let out for a short exercise. Before they return, place an older cock, which had previously been paired with the hen, in the box with her. Allow the young cocks to take stock of the situation, then put them in the basket, ready for the clubrooms and the racing preparations. They will be keen to get back from the race point and win back the affections of the hen. Of course they should be well looked after during this process, feeding on at least 1 ounce (28 g) of mixture each day.

The same method could be employed with just two young cocks without the introduction of the older one, allowing one into the centre section with the hen, then taking him away after a few minutes and replacing him with the other. Beware of the cocks beating or molesting the hen as they can become very boisterous.

## Pairing up and mating

One of the problems confronting the beginner is the difficulty in recognising a pigeon's sex! Generally speaking, cocks are rougher birds, with more flesh around the beak and eye, and they tend to be bolder and more robust when handled. Their wattle is larger and more wrinkled and their call deeper. On occasions, however, I have encountered hens that could be mistaken for cocks, and although this is not a usual occurrence, it does happen more with some families than others. When mating, the cock is the noisier of the two, while with Mealy or Red pigeons, the cocks usually carry black flecks, unlike the hens, and it is something of a phenomenon that irrespective of what colour you mate to a Red Cheq, any Reds bred from her will be cocks.

During the winter months, I usually note down my intended matings but I am constantly changing these pairings before I am satisfied that the cock and hen selected will produce ideal youngsters. A word of warning here. If you are fortunate enough to have a pair which is breeding winners, don't break them up. Allow them to continue and if they have been racing, pull them out of the team and put them into your stock loft so that they can continue the good work. If you lose even one from a race, and there is always that possibility no matter how well you prepare them, you may also have lost a whole line of successful racers.

After selecting the birds you intend to mate, and setting a date for this to happen, you should encounter few problems. It will help to give the cock the same nest box as he had the previous year, so keep a diary of who goes where. This will help him to settle more easily. If on the natural system, late February or early March is soon enough for mating, and some fanciers with a romantic streak like to let the birds get together on February 14 – St Valentine's Day! Few birds of any species take to the nest before this, so follow nature's way. As I have already mentioned, birds specifically intended for the later, long-distance and cross-Channel events, should be mated later.

I usually choose a weekend for pairing-up, spending most of it in the loft, for even the best-laid plans go astray. It always seems that something, no matter how little, will go wrong at this important time, forcing a quick change of plan.

The loft also requires a little extra attention at this time. I take the removable fronts from the nest boxes, and have them thoroughly scrubbed with disinfectant, then washed with warm water with a little disinfectant added – Jeyes is the brand I prefer. I treat my nesting bowls in the same manner as they are of the old red clay type. The boxes and fronts, when dried out, can be painted with white emulsion and for the bottoms of the nest box floors I cut up lengths of corrugated cardboard on which I spread two or three handfuls of coarse, dry sand with some sawdust and a small amount of garden lime mixed in. It is essential that the sand is dry.

Birds will tend to collect all kinds of bits and pieces to make their own nests, even if you provide them with nest bowl.

The pair will usually lay two eggs, the first about ten days after mating, with the second coming two days later.

No other birds other than those intended for mating should be in this part of the loft. I then start the process by placing each cock in the nest box chosen for him, fastening up the nest box fronts to prevent their escape. After two or three hours of confinement in the box, I allow the cocks out for some seed and drink, one at a time. By now they are familiar with their boxes and will go out and in freely.

I then let them have a selected hen and close the pair in the box, keeping a close eye on their advances to each other. If the cock gets too boisterous or abusive, I put a small makeshift division – a brick would suffice – into the box to allow the hen an escape route if she wants to relieve herself of his attention. Sooner or later they will take to each other, and I then allow them out as a pair so that together they can discover and use the box entrance.

When they are settled, and so as to be absolutely sure of the parentage of any youngster hatched, keep the pair closed up in their box except for feeding and watering, until the first eggs are laid. As soon as you notice the hen become 'eggy', put the nest bowl into the box, first sprinkling a handful of sawdust over the bowl bottom, and dusting with an insect powder.

Even after this has been done you may notice that the birds follow the natural tendency to build their own nests, picking up all kinds of things, bits and pieces, to line their nests as wild birds do. I have seen birds carry 6 inch (15 cm) nails into the nest box, also pieces of string and steel wire, while others show absolutely no interest in fending for themselves, content to make do with what has been provided for them. Either way it is an interesting thing to observe.

Make sure that while the birds are fastened in the boxes, they have an ample supply of grit, remembering the important part it plays in their digestive process by continually grinding the corn the birds consume.

Both parents take it in turns to feed the newly-hatched youngsters.

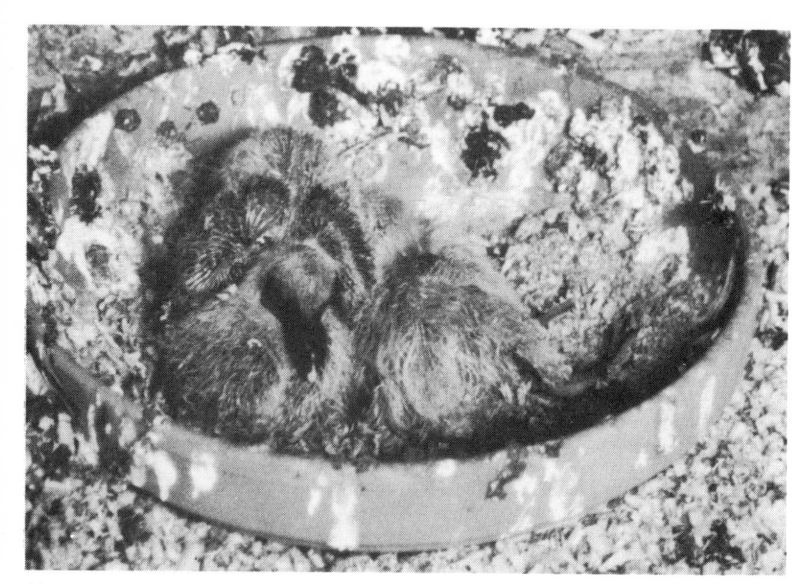

However tempting it may be, do not disturb newly-hatched youngsters as this will upset them.

The hens usually lay up approximately ten days after mating. The first egg is usually laid in the afternoon, for some reason between four and five o'clock, with the second egg coming two days later. The eggs are white, and about 1½ inches (35 mm) long. Both cock and hen share in the incubation, the hen usually taking over the hatching in the evening until morning, with the cock taking over around noon.

If all goes well and there are no 'clear' eggs, the hatching should take place within eighteen days and while birds are sitting they should be disturbed as little as possible. The newly-born pigeons are covered in soft, yellow down and are known as 'squeakers'. For the first few days the youngsters are blind, and the parents feed them with pigeon's milk, which is formed during the time it takes the eggs to hatch. It is interesting to watch the youngsters taking it by reaching into their parents' mouths with their tiny beaks. After about seven days, the squeakers continue to take milk, and also softened food, from the parents. By the age of three weeks they will have progressed to hard food. This is an exciting time for the loft manager as he observes them leaving the nest bowl to chase their parents for food. They will also be making their first serious attempts at picking up corn themselves.

During the breeding, do not stint with the food, or ration it to any extent. The feeding trough should always be well-filled at this time with a good mixture of grain. As an alternative, place gallipots in each nest box, preferably fastened on the outside as they can easily be overturned and their contents become contaminated.

Finally, don't forget to ring each youngster before the toes grow to a size that will make it difficult to slip on the ring.

At twenty-one days old, the youngsters can be taken away from their parents, and the hen will be laying again by that time and from that stage onwards they can be trained and managed as young birds, hopefully to provide you with plenty of future victories!

# BREEDING PRINCIPLES

This is a subject to which I have given much thought over many years and, unlike some aspects of the sport, it is not one which has experienced much recent change. Probably no subject apart from the long-ranging arguments over eye sign (the theory of judging a bird by its eye colourings and markings) has aroused so much discussion. Basically the question being asked is: should birds be 'out-crossed' or 'in-bred'. Out-crossing is the introduction or mating-up of a pigeon that is in no way related to its mate, while in-breeding is the pairing up of close relations, as near as brother to sister, mother to son or father to daughter. The principles of 'line breeding' are similar to in-breeding, with the matings still close.

All three systems have their advocates, but my own conclusion after years of reasearch is that in-breeding appears to be the quickest and best method of fixing type, although against this there is the point that if taken too far, the ensuing results could be disastrous as has been discovered by no less an authority than Charles Darwin. Darwin made an exhaustive study of the pigeon, finding in it a source of mystery and puzzlement, and one which he could not resist investigating.

His tests showed that continuous and uninterrupted in-breeding will inevitably lead to a reduction in size, strength, vigour and fertility of the offspring, and I have actually seen this occur in cases where fanciers chose to ignore the eminent naturalist. Indeed, some years back I purchased pigeons from a leading long-distance fancier who was a devout in-breeder, and one bird in particular, a blue late-bred was so fine and hen-like in its characteristics that I believed it to be a female bird for some time before discovering it was a cock. It did breed some good birds for me, but I was left agreeing with Darwin that in-breeding does produce smaller, finer birds.

Out-crossing, however, has the opposite effect and will increase the size and strength of the birds, although it does have the effect of breaking up the fixed blood lines. It therefore seems sensible to me to steer an intermediate course between the two extremes. The ideal way to do this could be to have one bird of the pair, preferably the cock, more or less in-bred and a typical example of the strain in which the breeder specialises, and the other a combination of two or more different families recognised for their long-distance achievements. In such cases the in-bred bird should predominate, and as a result the majority of the offspring should resemble the in-bred parent, yet also include the qualities of the other partner.

Whatever system of breeding is adopted, the breeder should not only concentrate his attention on the production of the good points, but also on the elimination of the poorer characteristics of the parents.

Any fancier with a really good in-bred family of pigeons, and of the opinion that he needs to introduce a cross, should make his new birds prove their worth before blending them into his own family. I have known many good racing fanciers whose plans and ambitions were set back for years through introducing birds that just didn't hit it off with the existing stock.

In-breeding tends to produce smaller, finer birds, while-out-crossing has the opposite effect. I have achieved the best results by steering an intermediate course between the two, mating an in-bred cock, typical of its strain, with a hen bred from a combination of two or more different families recognised for their long-distance achievements.

# AILMENTS AND INJURIES

Disease and injury is an occupational hazard for the racing pigeon, with the bird often the victim of stress, infection and a general lowering of resistance, although most modern racing pigeons are robust and healthy when housed in clean and hygienic conditions, and given good feeding, clean water and regular exercise. Their natural resistance to disease is high but you will be fortunate indeed – probably unique – if your loft escapes disease completely.

As with all living things, disease in a pigeon is an interruption of good health, a condition that every fancier wants his birds to be in. Happily, most diseases which affect pigeons can be successfully treated if detected early in the loft, but often the most effective way to prevent disease from spreading is to suppress the affected stock, distasteful though that may be. This would immediately check the unwanted intruder and a possible epidemic could be averted, so often you simply have to be cruel to be kind. However, do not forget that very often pigeons, especially young birds, can be cured of disease with no after-effects.

There are some excellent publications on the many forms of disease which can effect birds, and one on pigeon care and protection which you may find particularly helpful is available from Harkers of Lamberhurst, Kent.

What I want to draw your attention to is the need to control disease, and if possible prevent it from reaching your birds. Most important is loft hygiene. There is now sufficient evidence to prove that disease organisms, such as bacteria, viruses and moulds, accumulate naturally in the loft, causing only minimum damage until something occurs to lower the resistance of the birds, and then the troublemakers will make their presence felt. For instance, in the latter part of the year, particularly during the young bird racing season, respiratory diseases are at their very worst. If the birds do pick up an infection, and if you are alert enough to spot it early, these birds can be completely cured, but if ignored the bird's resistance will be affected and secondary invaders, probably viral, will then affect them, often leading to a long and tedious illness. Most lofts at one time or another have been visited by bronchial catarrh infections, a secondary invader that is very difficult to exterminate.

As the resistance of the bird can be suddenly and inexplicably weakened, it is only common sense to do your utmost to make sure that there is the absolute minimum of bacteria in your loft at all times. Cleaning and washing regularly with approved and specialised loft distinfectants is the only way of ensuring that disease-carrying organisms do not prosper.

Be especially careful with feeding and watering. What is the point of feeding with the best clean, sound corn, if the bird is going to pick the corn from a dirty floor or hopper because you provide it in a careless manner. Water, too, should be changed as often as possible, and before re-filling the drinker, make sure that it is extremely clean, always adding a sterilising agent, or even a little salt to the water.

If you keep your loft and utensils clean at all times, you will go a long way towards keeping diseases at bay. Be observant too – this bird's injured leg is obvious, but you can also see how it has fluffed out its feathers as if it is cold. Sick birds may also go off their food and often huddle in a corner. If you know your birds well, you will soon notice if any are unwell.

Another important way to prevent disease is to be observant. You will soon learn to recognise a healthy bird, and can then be on the lookout for any which are under par. In a healthy bird, the eyes should be clear and bright; the wattle clean and white and only greasy if the bird is feeding young; the beak should be clean, and the throat clear; the feet cold; and the feathers shiny and with plenty of bloom. A sick bird will go off its food and often huddle in a corner with its feathers fluffed out as if it is cold. Isolate any sick birds immediately and keep a careful watch on their progress.

These precautions must be taken, but there are diseases which may occur no matter how meticulous you are. Simple respiratory diseases, pigeon pox and coccidiosis can be transferred from a bird in one loft to another from a different loft. Pigeon pox at one time was virtually untreatable and any bird which had it was destroyed, but it is no longer necessary to do this, although if a secondary stage of the disease sets in there is often no other choice. Sores on the legs and the inside of the mouth, along with inflammation of the eyes, are 'pox' characteristics.

Coccidiosis is a serious, often a fatal, disease, with the bird at first appearing anaemic and droopy, and passing blood in its droppings. Usually only expert veterinary care will save a bird in this distressing condition. The problem for most fanciers is that such large numbers of birds are kept in each loft that it is impossible to engage professional treatment for each ailing one, and therefore most fanciers must often rely on commercially-available treatments and their own diagnosis. However, obtain professional advice where possible, especially if you feel there is something seriously wrong in your loft, as prompt attention and diagnosis often keeps losses to a minimum. In any case, never neglect that important first sign of disease or illness in a bird as it may be carrying something serious or infectious.

All fanciers must be strict about loft hygiene, as well as balanced feeding of stock. The overall effort required on the fancier's part is not too great, but the benefits to the pigeons are enormous in terms of lives saved and suffering eased.

### Treating injuries

I have seen birds return to their lofts with many kinds of injuries.

Sometimes when pigeons are lost in training or racing they are described simply as bad birds, but there could be many reasons for their failure to return.

Miraculously, they often manage to regain their lofts with terrible injuries caused in a variety of ways. By far one of the biggest dangers facing them in flight is the network of overhead telephone wires and electric cables, which they sometimes do not see and fly into. Such accidents usually prove fatal, and understandably there is a move afoot by fanciers in some parts to have these wires marked so that the pigeons could steer clear of them. The death rate among young pigeons when on the wing, either for training or racing, is sometimes colossal, and like most other fanciers I have had them alight on my roof with cuts and abrasions to the outer skin, broken legs and other injuries. These at first sight often appear worse than they actually are, and in fact I have often treated them myself to good effect.

If a bird has a flesh wound, cut away the feathers immediately surrounding it and bathe the injury in warm water with a small drop of iodine or TCP added. Gaping cuts will obviously require stitching and there are occasions when, if the bird is to survive, the only option is to do the job yourself. I have used a very fine sewing needle in an emergency, along with ordinary white spool thread, taking care and patience to knot each stitch separately and to perform the operation in scrupulously clean conditions. Take extra care to sterilise the needle, and before closing the bird up in an empty hamper, where it cannot be annoyed by others from the loft, sprinkle mild powder, such as baby powder, over the injury to keep it dry. I have seen birds of my own overcome frightening injuries after such treatment.

Broken legs are not difficult to set, and if the broken leg is not carrying the metal ring, so much the easier. Makeshift splints can be fashioned from light pieces of wood (I have used matches and lollipop sticks) as long as any sharp edges are rounded before using. Two or more splints are required, placed half and half over the area of the break. Bind them with tape or sticking plaster, first removing any feathers which are likely to interfere. Place the injured bird in a box, and to curtail unnecessary movement until the splints are removed, keep the food and water as near to the bird as possible.

# SOMETHING TO AIM FOR

One thing about pigeon racing – whatever you or your birds accomplish, rest assured that someone else has done better! Over the years many remarkable records have been set, and I'd like to draw your attention to a few – just to demonstrate what can be achieved by these marvellous flying machines, especially when the pre-race training and the conditions on the day are right.

Northern Ireland has long been regarded as a hotbed of racing, and the performance in 1961 of a bird owned by the Beattie Brothers stands to this day as a world record. It flew 186 miles (299 kilometres) from Dungarvan at an astonishing velocity of 2,857 yards per minute (2,612 metres per minute), or 97 miles per hour (156 kilometres per hour). Even allowing for a forceful tail wind, that speed over a distance of nearly 200 miles (320 kilometres) almost defies belief. One can only wonder at the reaction of the brothers when the bird zoomed into view less than two hours and five minutes after liberation.

Yet even higher speeds are possible, as proven by the Wickford partnership of A. Vidgeon and son when their most celebrated bird flew a 40-mile (64-kilometre) race at a velocity of 3,229 yards per minute (2,953 metres per minute) or approximately 115 miles per hour (185 kilometres per hour), in East Anglia in 1965. Rest assured that if one of your birds ever achieves such a speed, you'll certainly be among the winners at your club's annual prize-giving, and you may even secure a place in the all-time record books.

In marathon races pigeons record equally impressive feats, as demonstrated by R. and H. Kennedy of Irvine in Scotland when they won their country's National from Rennes in 1972. A fine performance this, for although weather conditions were not ideal for the entire journey, their bird flew the 534 miles (859 kilometres) at a velocity of 1,587 yards per minute (1,451 metres per minute), equivalent to approximately 50 miles per hour (80 kilometres per hour) while the bird was on the wing, which one can only assume was for every available daylight moment.

A year earlier, a bird had flown even faster than this for a significantly longer race. A 654-mile (1,052-kilometre) flight to Dorchester proved no problem for Mr C. Foster's winner of the Lerwick race, flying at a velocity of 1,603 yards per minute (1,466 metres per minute).

When we look at longer distances still, it is then that we see performances bordering on the incredible, and in such circumstances the true courage and determination of the racing pigeon has been well illustrated. Way back in 1913 the late C. H. Hudson, a famous Derbyshire fancier, sent his birds to Rome for a special 1,000-mile (1,609-kilometre) race. The birds were liberated at 4.30 a.m. Greenwich Mean Time on 29 June, and the fancier understandably presumed that all of his had been lost when they were still away a week later. Imagine his delight and surprise when, exactly one month after the liberation, a weary bird made its way back to his loft, just as he was sitting down to lunch. What a tribute to the character of that particular bird, to fly across land and sea, in changing weather conditions, for a full month and at the end of it all, alight at its home loft.

An Ayrshire fancier, W. G. Davidson, timed a bird in within nine days of liberation from Barcelona in 1960, a flight of 1,033 miles (1,662 kilometres). And in the United States, R. W. Taubert, a Texan, timed a bird which had flown across country from Manitoba, Canada, for 43 days and 10 hours, crossing mountain, desert and lakeland in its 2,000-mile (3,218-kilometre) mission.

To take things to what must be regarded as the absolute extreme, there is no more touching record of a pigeon's courage and dedication than that performed by a 'royal' bird more than a century and a half ago. The Duke of Wellington, a great admirer of the racing pigeon, decided to subject some of his best birds to the ultimate test and sent them to Ichabo Isle. You won't find that on any present-day maps, but for the record, it is situated in Western Africa. It is unbelievable to think that any of the birds ever returned to the Duke's London lofts, and to be absolutely accurate, none did.

But less than a mile (1.6 kilometres) from those lofts, one of the birds was discovered dead from exhaustion, 55 days after liberation. It had flown at least 5,400 miles (8,689 kilometres) before perishing, for while the line of flight is exactly that, it is likely that the bird would have detoured more than 1,000 miles

(1,609 kilometres) to avoid crossing the wastes of the Sahara, where it instinctively knew that neither food nor water would be available. Some estimates are that the gallant bird in fact flew some 7,000 miles (11,263 kilometres), averaging around 127 miles (204 kilometres) daily.

Some would say that to send a bird on such a mission is cruelty, but be that as it may, the statistics remain to this day, and surely serve as the ultimate measure alongside which the perseverence, stamina and strength of the racing pigeons of today can be measured.

It is qualities such as this which have endeared the racing pigeon to thousands of fanciers, and I would trust that by now you are among them. In fifty years of racing, I have had my pleasures and disappointments, but I have never lost my love of or admiration for these fine birds. They are capable of great things, and consequently I still have my many ambitions. I'll be happy if just a few of them are ever realised!

# INDEX

Numbers in italics refer to illustrations.